A Physicist Remembers

A Physicist Remembers

Richard J. Weiss

World Scientific

NEW JERSEY · LONDON · SINGAPORE · BEIJING · SHANGHAI · HONG KONG · TAIPEI · CHENNAI

Published by

World Scientific Publishing Co. Pte. Ltd.

5 Toh Tuck Link, Singapore 596224

USA office: 27 Warren Street, Suite 401-402, Hackensack, NJ 07601

UK office: 57 Shelton Street, Covent Garden, London WC2H 9HE

British Library Cataloguing-in-Publication Data
A catalogue record for this book is available from the British Library.

A PHYSICIST REMEMBERS

ISBN-13 978-981-270-058-2
ISBN-10 981-270-058-7

Printed in Singapore by World Scientific Printers (S) Pte Ltd

PREFACE

A funny thing happened to me on my way to a career in physics. World War II intervened and dumped me into the engine room of an aircraft carrier. Physics surrounded me — thermodynamics, kinetic energy, turbulence and mechanics were all there but no one was on board to discuss these subjects. The famous physicists were at Los Alamos. I was but a neophyte in the ship's bowels.

After Hiroshima my friends queried me about physics. They hadn't realized they were acquainted with such a smart fellow! I emerged as a wartime hero!

Richard J. Weiss

CONTENTS

INTRODUCTION

This scientific autobiography of Richard J. Weiss, protégé of MIT's John Slater and Bert Warren, traces the history of the 40-year effort to verify the predictions of quantum mechanics in describing the positions, momenta, and spin density of electrons on atoms. From this effort emerged the continuing worldwide Sagamore conferences to compare theoretical and experimental efforts in this field.

The research program began in 1950 when John Slater and four of his students joined the staff at Brookhaven and initiated Dr. Weiss into the intricacies of calculating electron distributions. With Slater's encouragement Weiss began an experimental program at Watertown Arsenal that eventually encompassed over 1000 physicists from most of the world intent on calculating and measuring the positions and momenta of electrons in crystals.

THE EARLY YEARS

In 1926 a three-year-old lad, familiarly called Richie, of natural curiosity and living on the ground floor of a two family wood frame house on Teller Avenue in the Bronx, was invited upstairs to watch the older boys play with their electric trains. The tenants that occupied the second floor consisted of one teenager Sammy who tolerated the youngster's presence and three adults who scarcely took note of him. While Sammy's attention was momentarily diverted Richie sat on the tracks and was shocked into a fit of crying when he discovered what Benjamin Franklin had known two centuries earlier, i.e. urine was an excellent electrical conductor. Richie's entry into the field of physics was ass backwards.

Richie's first memorable encounter with gravitational attraction occurred but a few years later when his father's model T Ford was parked on the hill outside his home. The six-year old would sit in the vehicle, wiggle the steering wheel, and fantasize a journey along a highway. When he accidentally released the parking brake and the vehicle rolled down the hill, harmlessly stopping at the curbside, he had sense enough to jump clear but he was not clever enough to avoid the thrashing meted out by his father. His rear end was impacted by Newton's fourth law of motion — every irresponsible action had an equal and painful reaction. Between Franklin and Newton and a smarting derrière, Richie began to feel inversely dynamic about physics.

His mother referred to her son's behavior as precocious. Not true — Richie was a brat. At his aunt's wedding he brought confusion to the ceremony when he chased a black cat up the church aisle and then messed up the reception that followed by vomiting on the dance floor. He increased the local entropy well beyond that governed by the second law of thermodynamics. He should have been thrashed outright at the time but the solemnity of the occasion spared him.

In the year he was born the Yankee Stadium opened a few miles from where his family took up residence in the Bronx. Richie was but five when his father took him to see Babe Ruth. The Babe could steal the show in right field by diving forward to catch a low fly ball and then rolling over several times until his forward momentum was dissipated. This dramatic display of the interchange of linear and angular momentum matched the crowd's reaction to the Babe's dynamic appearances at the bat. From that day on baseball assumed a higher priority for Richie than intellectual pursuits.

The large density of youth in Richie's Bronx neighborhood spawned a myriad of games, stickball being one of the more popular. It required a street, everyone had one, a tennis ball or a plain rubber ball — somehow one could always be found — and a broomstick. This last item was more of a problem. The boys frequently approached their mothers with little success. "Mom, do you have any old brooms?" only bought negative replies. The lads broke or lost these broom handles faster than the broom heads wore out and many perfectly good household brooms had been sacrificed in the name of good fellowship. Richie's grandmother took to locking them in a closet — they still had a short half-life since closet doors had to be opened occasionally. But, in spite of these sporadic displays of domestic larceny, Richie was still considered by his mother to be a 'good' boy, although he never remembered using a broom in the manner intended by the manufacturer.

To further demonstrate his unpuritanical behavior during those youthful days, Richie's ire was aroused when the nearby baseball field was closed to hardball (too many windows were being broken). A sign was painted to notify the neighborhood youth of this restriction. Richie and a friend found a can of white paint and covered over the sign one midnight. Alas! The paint was merely whitewash and the rain erased their misdeed, exposing the original warning. Regrettably, softball forever replaced hardball on that field. Since softball was not the national pastime the event presented Richie with an unforgettable period of deprivation and accompanying mental anguish. He and his friends would have to journey miles to find a grassy playing field.

Broken windows were not the sole peril in playing hardball in a populated neighborhood. On one occasion Richie's mother was walking on the sidewalk adjourning the ball field when a line drive struck her on the right side above the waist. She went down and we all ran to her. Almost unfazed she got up and assured us she was all right.

"Richie, your mother is the iron lady," remarked his friend.

He later learned his mother was wearing her armor plate. When Richie worked for the Army Material's Lab after the war he became interested in defeating projectiles and realized why his mother was unscathed in the incident. Her armor was called a girdle and it distributed the stresses over a large area. Physics will emerge in the darndest situations.

Acoustics played a role in Richie's youth. Mrs. Weiss had the best set of lungs on Teller Ave. When she called out "Richie!" a few times it traveled for several blocks and he'd come a running. His father had a subtler approach by whistling but two notes, an E with glissando up to a B. Why this carried as far as his mother's cry still remains a mystery. There is some interesting physics and physiology there.

Yet, scientifically naïve as he was, the laws of physics were ever ready to taunt him. A year before the stock market crash of '29, when the economy was still healthy, the family passed the summer on Coney Island. In spite of his mother's warning Richie spent the entire first day playing in the sand and running about on the beach, thus developing an excruciatingly painful sunburn. The UV portion of the sun's spectrum had done its damage. It was so painful that during his waking hours Richie took to sitting motionless in the kitchen. A woman visitor, noting his silence and immobility remarked, "Isn't he a good boy!" Even Richie was aware of the irony.

The huge three and four stack ocean liners that passed Coney Island on their way to and from their Manhattan piers fascinated the lad. It may very well have influenced his later decision to join the Navy and to write a novel about the Lusitania. Such ships as the Majestic, the Leviathan, the Berengaria, and the Mauretania conjured up visions of adventure and enormity.

Richie's further awareness of the perils of gravity and sudden changes in momentum brought him his first life-threatening encounter when the sled on which he was 'belly whopping' down a hill skidded on a patch of ice and caused him to ram his midsection against a tree. The diminished friction between a steel runner and the ice landed him in the Morrisania Hospital for several weeks. The damage to his liver was agonizingly painful but fortunately self-healing. As he lay in the hospital visions of death haunted him and he wondered what he might have become if he had been given the chance to grow up. He never guessed he'd be a physicist, or a scientist, least of all an 'unpublished' writer.

Before WWII physics was a remote discipline that not one in a hundred fathomed, the other 99 assuming the word to be a generic term for Ex Lax or milk of magnesia. Save for Albert Einstein, a household name for genius, and the Nobel Prize, a household expression for overnight fame and riches, physics and its bare fundamentals were not formally disclosed to Richard until he entered Morris High School at age 16. The early entrance into secondary school resulted from his having been advanced a full grade thrice in elementary school. This was affected when the Assistant Principal Mr. Price decided to redistribute the teaching load by fingering the better students for instant promotion. He entered class in mid term, pointed at several students, and uttered, "you, you, you, you, and you — leave your books and follow me." It was all over in less than a minute. In those days Assistant Principals were intimidating and adept at getting things done. (Richie rarely saw the Principal who he later learned was a serious tippler, preferring to have his assistant do the work).

Only a week prior to one of those on-the-spot promotions Richie was ordered to bring his mother to see Mr. Price. He knew why. There was a redhead named Virginia in his class on whom he had a crush but could never garner the courage to speak to. Passing her on the way home she smiled at him. He was totally tongue-tied and socked her to relieve the stress. Faced with the acid stares of Mr. Price, the girl and her mother, Richie found the words to apologize and from then on became a good friend of Virginia. She later entered a nunnery.

Experience became the young Bronxite's experimental method — like his first day in kindergarten when he ran into the little girls' room, quickly followed by the teacher who grabbed him by the ear and booted him into the boys' room. With only one bathroom at home he was too young to understand the nature of his 'misdemeanor' but in that brief moment at age five his innocence was wrested from him. He never did it again, the mark of a good experimentalist.

There was a mathematics teacher, Mr. Freeman, who tolerated no undisciplined behavior, rewarding the offender by grabbing his ear and pulling him out to the hall and leaving him there to face the stares of any who passed. Richie discovered that it was impossible to escape a firm hold on one's ear. Mr. Freeman should have realized that it was just such treatment by the conductor on a train that caused Thomas Edison's deafness when still a teenager. Richard's poor hearing would later surface whenever his mother asked him to go on an errand. Otherwise, Richard did well in Mr. Freeman's mathematics class; it was his low pain threshold around the ear lobes that received poor grades.

If anything inculcated in Richard an intellectual curiosity about the world, it had to be the hand-me-down set of 'The Book of Knowledge' that appeared in the house one day. Suddenly his life of baseball, stickball, basketball, and combating the Depression became enriched with cerebral matters such as mathematical puzzles. There were rainy days and the family didn't own a radio so 'The Book of Knowledge' was an awakening to a lore never discussed on the baseball diamond or while hanging around the local candy store.

The poverty of the early 30's left a deep imprint on Richard's psyche. Those days in the 30's, before Franklin Roosevelt and World War II turned the economy around, found many like his father feeling that capitalism had failed them. His father spent years looking for steady employment and was too proud to accept the dole. Richard would be sent miles to buy day-old bags of Wonder bread and would return embarrassed when the neighbors spied him. In those days, before preservatives were added, stale rolls were a bit like concrete and only dunking in hot coffee made them palatable. Left-wing sympathizers spewed out verbiage during the Depression that captured Richard's imagination. His mother and grandmother, both devout, were brought to tears when he repeated some of the anti-establishment 'party line' he later heard in college.

But in spite of hard times his parents shielded him and his younger brother from its bitterness. His youth was immersed in athletics and, like all normal boys, he dreamed of a life as a baseball player. His close relatives suggested that Richie should have his sights on school teaching, a well-paying profession during the Depression. Attending Junior High School he found himself inspired by the teacher whose principal claim to fame was that the famous baseball player Hank Greenberg used to be in his class and, pointing to where Richard sat, identified the very desk. A classmate of Richard's, Gabriel Pressman was outstanding in English — his compositions demonstrated a professional command of the language far beyond that of anyone else in the class. Yet Richard still felt smug knowing that Gabriel couldn't throw a baseball. Furthermore who wanted a name like Gabriel? Pressman later became a successful reporter whose name would be frequently accredited with a story — who could forget a name like Gabriel Pressman? Still, the ghost of Hank Greenberg made a far greater impression on Richard. He was exuberant when it was announced in his senior year at high school that a baseball team was to be formed. Richard tried out but the coach took him aside and informed him that he was a good ballplayer but was too young for the team — no fault of his own but he was shattered — blame Mr. Price for the promotions.

In his first formal encounter with physics at High School Richard was scarcely impressed with the contents of the course although he still maintains a faint recollection of the instructor's teaching technique. If it wasn't in the textbook it wasn't right. Richard concluded that the teacher really didn't understand the subject but was smart enough to keep one chapter ahead of the class. What a pity that the physics of baseball was not part of the course — that would have struck out two birds with one pitch. Physics teachers today could well benefit by approaching the subject from home plate.

As a senior in high school Richard felt trapped — there was no money for college and the job market was dismal in 1939. His father felt helpless in not being able to send Richard to college with a flowing green campus. Fortunately New York City boasted CCNY, a free school for the academically gifted. So for a registration fee of $2 he enrolled in that crowded institution, a school that many years later even hired a physicist, Robert Marshak, as President. Having learned in later years that General Colin Powell attended both Morris High School and City College, Richard wrote to him and was pleased with his reassuring reply about the quality of education he had received at that college. He probably never sat at Hank Greenberg's desk, thought Richard, for then he may have achieved fame as an athlete rather than as a military genius!

For two years Richard boarded the 5¢ subway on The Grand Concourse at 170th Street and rode to mid-Manhattan where he attended crowded classes at the campus on 135th Street and Convent Ave. Walking up 135th Street one morning he spied a truck carrying tanks of bottled gas. The vehicle hit a bump and one of the tanks fell off causing the valve mechanism to rupture. The tank careened rapidly and spun aimlessly on the street as Newton's third law of motion came into play from the escaping gas. Richard ran for cover until the 2000 pounds of pressure had been reduced to naught. The sight and sound of such a catastrophe was unforgettable — it was a physics demonstration that presaged the rocket age a decade before its time.

Richard was a terrible student, spending most of his time practicing with the CCNY (Beavers) baseball team. But at the end of his second year he had to declare his specialization to the academic authorities since degrees were not granted in baseball. Mathematician? He wasn't good enough. Chemist? The laboratory work was too sloppy for him. Biologist? He hated dissection. Physicist? Yes — it was the only scientific subject left and the least messy. Not surprisingly then, he again backed into the subject.

As a junior at CCNY he became a regular on the varsity baseball team, probably proving that he was a better baseball player than budding physicist or that the gifted kids that attended CCNY were not too well coordinated. The lacrosse team also practiced at Lewisohn Stadium under the tutelage of 'Chief' Charles Bender who took Richard aside one day and suggested he take up lacrosse, "A better game than baseball," the 'Chief' declared. Surprisingly enough the Chippewa native American 'Chief' Bender had been an outstanding pitcher for the Philadelphia Athletics from 1903 to 1914 and was elected to the Hall of Fame at Cooperstown in 1953. In 1942 Richard had never heard of him.

While some of the outstanding physicists in the world were beginning to concern themselves with uranium fission, and while the war in Europe drove most of that subject underground, Richard wrestled with the Introduction to Thermodynamics course given by Zemansky and with Atomic Physics delivered by Semat. The former course turned out to be almost totally devoted to mathematical manipulation of equations, the real world of thermodynamics escaped Richie's comprehension. Even boiling a pot of water to make tea enflowered no new romanticism or scientific significance from Zemansky's equations. Physics began to lose its appeal, if, indeed, there ever was any. Almost 40 years later Richard suddenly 'discovered' the subject of thermodynamics when Clarence Zener pointed out that the cusp in the specific heat of iron metal could be related to its magnetism. After the Zener lecture Richard remarked that at long last thermodynamics had displayed some usefulness! He went on to publish extensively on that subject.

It was early in 1941 when Richard's father bought a diner in Brooklyn and father and son would spend weekends running the establishment. With not an iota of help from thermodynamics Richard learned the rudiments of short order cooking. It was one Sunday in December, while clearing the counter, that he heard about Pearl Harbor over the radio, and less than a year later Richard, still bearing memories of those ocean liners passing Coney Island, signed into the Navy V-7 program, permitting him to remain in college until the Navy needed him.

The summer of '42 approached with America gearing up to a wartime economy. In former years Richard had managed to find menial jobs during those summer holidays but this time he got lucky. The State unemployment office sent him to a large unmarked building on Seventh Ave. in downtown Manhattan where he was offered a job as a custodian. Half an hour later he found himself sweeping the floors of the wartime Office of Censorship, a multistory building and former mattress factory with thousands of employees reading and excising foreign mail.

He asked his supervisor, who took a shine to the lad, if there might be something else he could do. The only thing he really knew about brooms was how to saw off their handles for stickball. When he bragged that he was good in mathematics he was reassigned to the Leave Department, calculating by hand the amount of annual and sick leave employees earned, and deducting what had been used by each worker. At least half the employees, expert in their own language, presented a problem for Weiss in trying to explain in English the somewhat complicated system for amassing and using leave. Furthermore, he was seated near the door leading to the Ladies' room and the passing pulchritude imposed an added burden in trying to concentrate on his math. Diplomacy was also required. Employees could only be credited with sick leave if they presented Richard a note from a doctor. An order from on high denied sick leave for all dental visits. When Weiss tried to explain this to someone who had suffered an impacted tooth but spoke only an African dialect his resource-fulness was tested to its limit.

Unclaimed paychecks due to illness were left with Weiss with instructions to only hand them out with proper identification.

One individual had left his badge at his workstation but showed Weiss his last two unnegotiated paychecks! The exercise in explaining the mathematics of leave recording to one unfamiliar with English awakened in Weiss a few adult qualities such as sympathy and patience.

On July 1, 1943 Richard was called to active Navy duty, bidding goodbye to the CCNY baseball team and his fellow stickballers on Teller Ave. His orders specified a semester at St. Lawrence University, a school that claimed Kirk Douglas as one of its alumni.

Richard scarcely remembers the goodbyes he tendered his family when he left for St. Lawrence although he does recall meeting with dozens of other Navy enrollees at Grand Central Station for the day-long train journey to Canton in upper New York State, the furthest he'd ever been away from home. The female-male ratio at the college had been increasing since the beginning of WWII so that the sudden appearance of over a thousand sailors elated the coeds as well as the tars. The underage Seaman Weiss got to know nary a one (Assistant Principal Price take note), his spare time was devoted to playing the bass drum in the marching band, learning to swim in the nearby river, and cramming for exams.

The bass drum presented a challenge since the drummer was responsible for maintaining the marching beat, 120 booms per minute. Needless to say the weight of the instrument produced a steadily tiring effect and a resulting slowing down of the beat. When the Admiral arrived after two months for a grand review of the regiment the long column of sailors found themselves hopelessly out of step with the drummer. Fortunately for seaman Weiss there was no Navy regulation by which a drummer could be court marshaled for an unsteady beat. An erratic frequency was added to Richard's misadventures with the subject of physics.

Towards the end of term the school announced a dance but no band was available. In a short span of a week seaman Weiss had discovered enough talent amongst the sailors to put together a respectable ensemble of piano, trombone, trumpet, drums, clarinet, and his own guitar, which his father had brought up from the Bronx on a weekend visit. Alas! The Saturday night shindig faced disaster when the piano player broke his wrist at football that afternoon. Without a piano every member of the band quit — a human chain reaction. How seaman Weiss managed to find two accordions to add to his guitar he scarcely remembers. Visions of rebuke from the commanding officer haunted him all afternoon but somehow there were enough decibels emitted by the three pieces to satisfy the dancers. Weiss was even complimented for the effort — everyone knew that one made sacrifices in wartime.

St. Lawrence was followed by a '90-day-wonder' course in marine engineering at Annapolis. Richard had actually requested an assignment in physics but the officer in charge couldn't accept the notion that physics and marine engineering differed. During his three months as a midshipman at the Naval Academy Richard was immersed in regurgitating Naval engineering regulations for the weekly exams. He frankly understood nary a thing about the engines that propelled a ship and any physics he had retained from his CCNY days was by now virtually forgotten. He managed, though, to survive the Annapolis military discipline and on April 26, 1944, not having reached his 21st birthday, a young ensign was sworn in at the Naval Academy and assigned to engineering duty on the CVE 100, a baby aircraft carrier being commissioned on the west coast. When he left for Seattle to join his ship Ensign Weiss felt important, the conviction of being an officer and a gentleman deeply entrenched in his psyche. But on the flight to the west coast he became airsick and repeated his performance at his aunt's wedding — he made a mess of his confidence and his uniform. How rapidly the high and mighty can fall, he thought.

To one who had never been on any marine vessels other than the Staten Island ferry and a rowboat in Central Park, the baby carrier was grandiose. Accorded a private stateroom the new ensign was elated. He even volunteered to play the portable organ for church service and was permitted to keep the instrument in his cabin. But placed in charge of the main engines because of his rank presented a sticky wicket! It took Ensign Weiss a year to fathom what most seamen had learned from long service in the Navy, i.e. how to use a monkey wrench, how to read a pressure gauge, and how to avoid seasickness. If Zemansky's course on thermodynamics had any meaningful relationship to the reciprocating steam engines that propelled the aircraft carrier, it failed to make any impression on the young ensign that had just reached his majority. Yet, the faith that Assistant Principal Price had in accelerating Richie's education was at last coming to fruition.

After a year his understanding of the functions of the machinery on board ship began to surpass that of most seamen. He discovered amongst the ship's assigned paraphernalia an indicator gauge — a small device that recorded steam pressure versus time within the main propulsion cylinders. He managed to record this information but Zemansky hadn't given him a clue what to do with it! Perhaps physics was too deeply imbedded into his subconscious to be useful!

His principal memories of those war years were boredom and fatigue. Floating around the endless South Pacific, the sea varying slightly from day to day as it responded to the winds, and the flying fish occasionally soaring thirty feet, Ensign Weiss found little to commend the scenery. Even standing well off shore during the invasions of Iwo Jima and Okinawa seemed like just another boring day. The pedal driven organ provided some relaxation although it hardly enamored him to his fellow officers trying to catch some shuteye during the day. That entire expanse of the ocean seemed like one huge bath tub, the hourly recorded readings of 85° Fahrenheit never varying for weeks at a time nor the 100° engine room temperature where he stood watch. Weiss witnessed windless days when nary a ripple appeared on the sea, although on one occasion when the entire fleet sailed through the eye of a typhoon off Okinawa he looked up at waves at least 150 feet in height. The aerologist (weather man) reported record winds over 150 miles per hour. The morning after the storm the entire five dozen place settings laid out the prior evening for dinner in the wardroom were haphazardly strewn in one corner.

Only one picture of man's might stands foremost in Richard's wartime memory. When the 8th fleet amassed before the Okinawa invasion there were men-of-war as far as the eye could see. Perhaps 500 ships of every description dotted the skyline. But what a way to combat boredom — two years of ennui for one day of spectacle!

There was another memorable episode that relieved the tedium. The carrier developed a list of several degrees, perfectly normal as oil and ballast are respectively used up and taken on. While on duty in the engine room the OOD called down to advise him of the list and asked how long it would take to bring the ship back to an even keel. Weiss conferred with the appropriate seaman and advised the bridge it would be completed in two hours. It actually required five hours and Weiss was ordered to the bridge where he was given a dressing down by the captain.

"Don't you realize that there is a war?"

"Yes, sir."

"Orders must be carried out with precision."

"Yes, sir!"

"This could lead to a court martial. Understand?"

"Yes, sir."

"Dismissed."

He later learned that the captain had emerged from his cabin at the start of Weiss' watch for his daily sunbath but the ship's list had caused the shadow of the bridge to shade his favorite spot. He returned in two hours and 'hit the roof' at Weiss' failure to bring back his sunshine.

Worse than boredom was the fatigue experienced in the forward combat area when duty cycles of four hours on, four hours off, completely screwed up his circadian rhythm. His very being revolted at this abuse of his natural biology. One never managed enough sleep!

The movies shown every evening on the hangar deck did relieve the monotony and on one occasion a destroyer providing escort in case of submarines challenged the carrier to a game of basketball. In no time their team was hauled on board via breeches buoy and the canteen was opened to sell ice cream, cigars, and sweets to the cheering spectators. Still, boredom, lethargy, listlessness — call it what you wish, its only conquest appeared to be booze and broads. All officers aboard ship were provided with a small combination safe to store their classified papers. Richard's safe contained two illegal bottles of Johnny Walker and a service pistol that he was too frightened to use. One bottle of scotch disappeared during the basketball game but no broads, not even a few mermaids could be netted as cheerleaders.

Richard's baseball passion was aroused when Johnny Mize, Hall of Fame first baseman for the New York Yankees, joined the ship as an athletic director. Richard and 'Johnny' spent hours talking about baseball — what else was there to do? This encounter even surpassed the elation Richard shared in his tenancy of Hank Greenberg's desk.

Richard's division had an old time warrant officer, recalled to service after Pearl Harbor. He suggested to Richard that they have some target practice with their 45 caliber pistols during a few idle hours in San Diego. And so waving off ear plugs and never having fired a pistol Richard shot off hundreds of rounds of ammo and spent three days stone deaf, five days half deaf, and several weeks with a slight buzz in the head. That presented an interesting bit of psychology cum physiology in demonstrating the ability of the body to ignore certain stimuli when experienced in excess. Modern high decibel stereos have been known to permanently impair hearing but in Richard's case subsequent hearing tests years later revealed no impairment from that one encounter. For a time, though, Mr. Freeman had his revenge!

When the war ended with the atomic bomb dropped at Nagasaki physics was skyrocketed to its zenith of importance. Suddenly his family and friends took a second look at young Richie, equating him with Oppenheimer and Lawrence. He was questioned about physics, the bomb, neutrons, etc. He actually had scant knowledge but he had found it easy to answer with authority — like his high school physics teacher he was one chapter ahead. Baseball was soon forgotten.

In 1946 the G.I. Bill appeared as a gift from the sea and Ensign Weiss registered for graduate work in physics at the University of California at Berkeley, a school with such distinguished staff members as Oppenheimer, Lawrence, McMillan, Segré, Serber, etc. Suddenly his ignorance of physics struck him between the eyes when he attended his first classes. It became a long, hard pull to catch up with the mainstream of students and to convince his academic advisor, Francis Jenkins, that he belonged at Berkeley. Indeed, after two years of hard work he had made considerable impact on the subject and even courted original research. This part of the story is worth repeating.

It all started when Grant Fowles, later a professor of physics in Salt Lake City, and Weiss teamed up to perform some spectroscopy for Francis Jenkins' laboratory course. Fowles had read an item in Richtmyer's book on modern physics that cited an experiment by Williams at King's College London (later to have Richard for ten years as a visiting Professor) in which he noticed an anomaly in the spectrum of hydrogen — one of the lines appeared to be split, but only barely. Fowles correctly deduced that the splitting for helium would be 16 times greater and therefore easier to observe — except for one thing. The line was in the far ultraviolet and the entire experiment had to be performed in a good vacuum. This technology wasn't simple for young graduate students in 1946, even for one who knew how to wield a monkey wrench. The two graduate students approached Professor Serber, theoretical leader at the former Manhattan Project, asking his opinion whether the Williams' experiment could be correct. He shook his head and assured them that quantum mechanics was accurate. The pair remained undeterred and went ahead. Years later when Weiss ran into Serber and asked about the discovery, he received but a shrug of Serber's shoulder. Physics is always full of surprises.

The famous spectroscopist R.W. Wood visited Berkeley. He had fabricated the diffraction grating employed by Fowles and Weiss in the experiment and related a few humorous and bawdy stories for Fowles and Weiss' benefit. Only years later did Weiss learn that Wood was renowned as a showman. His book *HOW TO TELL THE BIRDS FROM THE FLOWERS AND OTHER WOODCUTS* is a classic.

While Weiss was grappling with physics he still bore in mind the prewar encouragement he had received about teaching. He enrolled in the sequence of classes given by the Education Department for a teaching certificate and terminating with a term of practice teaching at Oakland High School. The unruliness of those teenagers convinced him that teaching would be an unwise choice and he never considered it again.

After a number of false starts the experimental results with helium became clear — the spectral line was indeed split! What a triumph for two grad students! Alas! Two weeks later Professor Willis Lamb at Columbia reported in the Physical Review the same observation but measured in a more elegant fashion. He had beaten them to it! It took some years for theoreticians in physics to explain that the effect was due to an unexpected behavior in the electron's magnetic moment. Lamb's discovery received the Nobel Prize in 1955.

In those days Professor Oppenheimer was the physicist's superman. The gaunt cadaverous figure had lost considerable weight during his Los Alamos days but his smooth delivery mesmerized Weiss. 'Oppie' could convince you that you understood his exposition — only when you tried to work out the rationale for yourself did the unanswered questions cloud the issue. Oppenheimer chain-smoked, frequently interchanging chalk and cigarette. He deftly avoided placing the chalk in his mouth or writing with the cigarette.

Segré was the opposite. His Italian accent, spectacles, and informal demeanor, made him look and act like everyone's uncle. Richard walked away from Segré's lectures with clear pictures about nuclear physics. Between lectures by McMillan, Seaborg, and invited speakers like Felix Bloch the entire panoply of famous characters in the era of nuclear physics left Richard with distinct impressions as to how one spoke and acted if one were to be part of this elite physics community. Funny, Richard thought, none of the lecturers ever mentioned what they didn't know.

In 1947 Richard decided to return to New York with his University of California Master's Degree and to complete his graduate work at Brookhaven, a new laboratory on Long Island dedicated to the nuclear physics that emerged with the bomb. At Berkeley he had sat in on courses given by those associated with the frontiers of science in reactors, radioactivity, and cyclotrons. Brookhaven would afford him the opportunity to actually work on an atomic reactor.

Physics was now equated with glamour. Physicists were viewed in society as the golden-haired boys — they were patently geniuses. Whenever someone discovered Richard was a physicist they immediately apologized for their lack of knowledge of the subject. "That's alright," Richard assured them, "I won't quiz you."

To save money on the journey east Richard joined a car pool, the other passengers expecting him to share the driving. The last disastrous time Richard held the wheel of a car was in a Model T Ford parked on a hill in front of his home in the Bronx. His youth and the rush of events had deprived him of the opportunity to learn to drive. The Navy had taught him to run the main engines on an aircraft carrier but not to drive a car. After all, Einstein had learned to sail a boat but did not drive — Richard was in excellent company. The other passengers in the car pool were dumbfounded to learn that Richard was a non-driving nuclear physicist. Queer ducks these physicists!

The journey through Washington's apple country, Montana's Glacier National Park, the endless plains of the Midwest, and the smoke of Pennsylvania, provided a striking contrast to Richard's life of stickball on Teller Ave. and his endless views of the South Pacific. Before entering the Navy he neither smoked nor drank. But scotch and cigars came easily. He still remembers the drinking fountain just outside the main bar in Canton, NY, home of St. Lawrence University. Carved on the granite bowl was:

GIFT OF THE W.C.T.U.

The Women's Christian Temperance Union may still exist, like teetotalling Baptists, but Benjamin Franklin provided ample proof that man evolved to drink. The position of the elbow is precisely placed to bring the goblet to the lips. If closer to the wrist the hand would not reach, while if closer to the shoulder it would overshoot the mark. 'Poor Richard' Weiss had been lured into following the precepts of America's first physicist, never dreaming that someday he'd run a colonial tavern circa 1780.

BROOKHAVEN

John Pasta, Ira Bernstein, and Richard Weiss journeyed en trio to Brookhaven National Laboratory in 1948, the first fellows appointed to that position from New York University. They scarcely knew what to expect. The three had completed all the formal course work for the PhD — only the thesis requirement remained. Brookhaven offered them a stipend, a refurbished apartment each, and assigned them each an advisor from the physics department. Weiss' mentor turned out to be Simon Pasternak, later editor of the Physical Review.

Prof. Joe Boyce, chairman of the NYU physics department was Weiss' academic advisor. One empathized with Prof. Boyce — to lure him to NYU he had been promised a new physics building to replace their antiquated Victorian structure on the Bronx campus. After a ten-year wait he gave up and tendered his resignation. A week later the building burned to the ground presaging the eventual closure of the campus and transferring the college to Washington Square in lower Manhattan. Boyce confided to Weiss that if he had delayed submitting his resignation by just that one week he would have looked like a rat deserting the ship and jokingly accused of 'arson'. Boyce happily moved to Argonne, one of the string of atomic energy labs that experienced rapid growth after the war.

The Brookhaven Laboratory, 73 miles from New York City, was established at the former Camp Upton site, which dates back to WWI and is especially remembered for Irving Berlin's musical 'Yip Yap Yaphank', named after a nearby town. The song from the show "Oh, How I Hate to Get Up in the Morning" is still played as a reminder of the Great War. After Pearl Harbor more wooden barracks had been added to handle the larger numbers of boot camp draftees. At the war's end the entire establishment was turned over to the Atomic Energy Commission (AEC) as a facility for atomic research, seventy three miles being considered a safe

distance from New York City in case of a nuclear mishap. The
facility's operation was supervised by ten eastern universities,
from Harvard to Pennsylvania. It was 'rumored' that the site was
chosen as equally inconvenient to all ten schools. It could be
reached in two hours by road from New York City and in a more
tiresome journey by the notorious Long Island RR at a period
when that train service was becoming world famous for its
unreliability.

On arrival at Brookhaven the three doctoral candidates met
Sam Goudsmit, famous for his theoretical work on the structure of
the electron and for his wartime mission as head of the ALSOS
project, a secret group under Goudsmit's scientific leadership that
arrived in Europe after D-Day to find out what the Germans were
up to in uranium research.

Weiss was assured by Tom Johnson, chairman of the
physics department, that the new atomic reactor would go critical
in a matter of weeks. Alas! The huge graphite assembly for
moderating (slowing down) the neutrons had been designed by a
group of physicists with minimal civil engineering experience.
Luckily, just before startup an outside consulting engineer
discovered that the structure was not properly supported. The
entire assembly had to be redesigned and rebuilt, a two-year
process. If it had gone critical it would have been too radioactive
to modify and might have produced the most scarifying and
expensive incident in the Atomic Energy Commission's short
history. Whatever the case — this left Richard up the creek without
a neutron.

But before engaging in any research at the reactor Weiss
had to submit to a 'Q' security clearance. Many buildings
including the reactor could only be entered by those with
appropriate sanitizing. This procedure was considered a necessary
nuisance by researchers and it occasionally created friction
between scientist and security officer; the former deemed their
basic research of no conceivable value to the Russians while
the latter merely carried out their orders. The entire adversarial
atmosphere between scientist and security personnel mushroomed

like an atomic bomb when Klaus Fuchs, Allan Nunn May, and the Rosenbergs were exposed; and Senator Joe McCarthy brought shivers to the populace by hinting that Russian spies and sympathizers lurked behind every bush.

The Fuchs case was particularly disturbing since he was a competent physicist who had worked on and become privy to most of the essential details of the Manhattan Project and had been befriended by many of the leading scientists in spite of his reputation as a 'loner'. He managed to transmit to the Russians details of bomb construction as well as the Hanford reactor 'blueprints'. Fuchs had had a trying time growing up in Nazi Germany. He had embraced communism as a post-Great War philosophy since it offered promise to bring order to a confused society, but after he suffered persecution by the Nazis he fled to England. Communism remained imbedded in his psyche. When confronted with his duplicity he meekly cooperated with the authorities. Sir Michael Perrin, director of the British atomic energy establishment received Fuchs' confession. Sir Michael informed Weiss in later years that Fuchs fully expected the death penalty, not realizing that Russia had been an ally at the time of his misdeeds.

After receiving his 'Q' clearance and completing the assembly of his experimental equipment Weiss needed a source of neutrons. Fortunately there were two active research reactors in the USA at the time, one at Oak Ridge, the other at Argonne. Through the kindness of Cliff Shull, soon to discover antiferromagnetism and later to become research professor at MIT and Nobel laureate Weiss was afforded space at X-10, the nomenclature for the Oak Ridge site.

And so it came to pass that Weiss, having learned to drive, loaded his experimental equipment on the AEC station wagon and journeyed to Oak Ridge with Andy McReynolds, a lanky Texan on the Brookhaven staff who specialized in solid-state physics. His slow southern drawl belied his shrewd understanding of physics and many under-rated his ability because of his measured delivery.

Andy was instructed to look after Weiss, "Make sure he doesn't kill himself!"

The soil of Oak Ridge, rich in red iron oxide and lacking rocks, could quickly turn into a mud that clung to everything. What a mess cleaning it from shoes and clothing! To Weiss who had met southerners on board ship during the war and found the issue of black segregation distasteful and adversarial, he realized he was now in the enemy camp. Fortunately he was too busy with physics to engage in any debate over social issues.

The subject of his thesis *SMALL ANGLE SCATTERING OF NEUTRONS* consisted of sending a well-collimated beam of slow neutrons through samples of tiny particles. Just as light is bent at an air-glass interface so one expected a similar behavior for neutrons, depending on the index of refraction and the size of the particle. In those early postwar days companies leaned over backwards to provide small samples of their wares to the AEC and a myriad of different powdered substances were collected gratis by Weiss. He went to work. After several months of eating southern ham and hominy grits, washing red mud from his clothes, and performing measurements on the scattering of the neutron beam, Weiss had collected enough data to compare with his own theoretical calculations and to grind out a thesis. What had he shown? The concept of an index of refraction worked for neutrons as well as for photons — no surprise. Still — someone had to demonstrate it. And so back to Brookhaven and civilization to write up the research, Cliff Shull's kindness forever to be remembered. Years later Weiss hinted to Professor Slater at MIT that Shull could be a welcome addition to the MIT reactor staff and the move came to fruition. Shull had some boys approaching college age and academic faculty was granted special privileges for their offspring.

Many of the world's illustrious physicists placed Brookhaven on their itinerary, Einstein the only exception Weiss could not recall meeting. Even the less illustrious came. Weiss first met Robert Maxwell in 1949 when he was preparing his thesis at Brookhaven. Maxwell promptly asked Weiss to write a book for his publishing firm Pergamon Press but Weiss skirted this request since he recognized his scientific limitations at that stage of his career.

The young Weiss couldn't help but be impressed with this tall, striking, good-looking figure who exuded authority and charm. He was just starting as a scientific publisher and was touring America to find authors. His stock in trade was his persuasiveness and a collection of titles that were confiscated from the Berlin publisher Springer Verlag under the Alien Property Laws.

Weiss recalls Maxwell's wife relating to him that only a year earlier Maxwell had been warned by his doctors that he had contracted a terminal illness and was advised to put his affairs in order. The nature of the misdiagnosis was never determined as Maxwell rapidly recovered. After that dinner meeting, which included some of the dignitaries from Brookhaven, Sam Goudsmit confided to Weiss that he had met many such unscrupulous entrepreneurs like Maxwell. That surprising remark deflated the man's image and Weiss always remembered it, its significance borne out by subsequent events.

Ten years later Pergamon Press was well ensconced in Fitzroy Square, London, where Maxwell built an empire with the simple discovery that libraries would automatically buy scientific tomes, no matter the quality of the writing. He was in the midst of shaking up an otherwise staid industry of English publishers with this philosophy. He was also experimenting with various name changes such as I.R. Maxwell, Capt. Robert Maxwell (he served with distinction in the army), Ian R. Maxwell, and finally plain Mr. Robert Maxwell.

When Harvey Brooks, a noted physics professor at Harvard, took on the position as Editor-in-Chief of a new Maxwell journal Weiss concluded that Pergamon Publishing had gained respectability and he submitted an article for the first issue. He also decided to take up Maxwell's ten-year-old offer to write a book and submitted the idea through Brooks. Shortly afterwards Weiss signed a contract with Pergamon Press to produce one on Solid

State Physics. In London for a year at Imperial College, Weiss visited the Fitzroy Square headquarters for Pergamon.

As soon as he entered the office Maxwell handed him a copy of his letterhead announcing his candidacy for Parliament under the Buckingham Labour Party banner. A handsome picture of the candidate adorned the sheet and he listed himself as Captain I.R. Maxwell, M.C., the final initials referring to his military decoration. He further advised Weiss that the Pergamon empire was outgrowing its London headquarters and he was moving to Headington Hill Hall in Oxford, a stately home that better suited his style and would enable him to live 'over the store'. Before Weiss left, Maxwell received a phone call and gave the caller 'what for' in one of the many foreign languages in which he was proficient. When he slammed the receiver down he turned to Weiss and said, "That's the only way to talk to a German." Maxwell's Jewish and Czech background probably colored his outlook.

As Weiss discussed the book's format with members of Maxwell's staff, the subject of Maxwell's personality often arose. Weiss was informed that Maxwell's announced candidacy for Parliament left most of Pergamon's employees with mixed emotions. If Captain Maxwell succeeded it would get him out of their hair but what would it mean to the country? Stories of Maxwell firing his entire staff on a Friday afternoon were rife. A colleague who often dealt with Pergamon told Weiss that Maxwell would pay his bills but withhold 5%, most creditors forgiving the self-imposed discount. If one made a loud and repeated complaint the remainder would eventually be remitted.

There is no question that Maxwell shook up the scientific publishing industry with his proliferation policy and the jury is still out as to whether the net result has been advantageous to the technical community. Weiss, for one, was certainly pleased with the layout and handling of his book.

A year after Weiss' book appeared he received a letter in French from a physics postdoc in Czechoslovakia (Maxwell's country of origin) who pleaded that it was impossible to secure the hard currency necessary to purchase the book. Weiss forwarded the letter to Maxwell who replied to Weiss that his stock of spare copies was exhausted and he regretted that he could not honor the request. A month later Weiss received a note from the postdoc in Prague thanking him for the copy of the book! Maxwell was unpredictable!

A decade passed before Weiss saw a picture of Maxwell in a London paper. By now he had grown considerably and was in the process of publishing aggrandizement, well beyond the field of science and in a direction which would eventually lead to his becoming newsworthy. He had completed his tour of duty as a member of Parliament, yet the country survived.

During the height of the cold war Maxwell published an information guide about the USSR (Russian was one of his accomplishments). It represented an effort to break down the barriers between East and West. The London Sunday Express of December 8, 1991 showed Maxwell in a huddle with Gorbachev and Howard Baker, all three enjoying a good laugh. The headlines indicated that Maxwell was responsible for arranging a Russian loan of several billion pounds and that Pierre Salinger was privy to the loan discussions.

Maxwell also became instrumental in getting the Russians to honor royalty agreements. Weiss was well aware of this problem. After Pergamon Press published his book a Russian publisher translated it without permission. All Weiss got was one presentation copy.

The final chapter in the Maxwell saga is with his mysterious drowning. Whatever his complex financial manipulations he has left his mark on scientific publishing as well as leaving the world in a dramatic fashion. He had a large family but only two of his sons seem to be left with the task of sorting out the fiscal machinations of their father. Having never worked for Maxwell Weiss found him pleasant to deal with. When one was in his presence one knew that things would get done.

With his thesis completed and accepted in 1950, Weiss' Brookhaven fellowship came to an end. He now had to face the world and obtain his first real job in an economy that had begun to tighten as the country retrenched from its wartime mission. Weiss was smitten by the research bug and wished to remain at Brookhaven doing neutron diffraction, now that the reactor was ready to go on line. The Brookhaven budget could not support another member on its research staff (so he was told) but Tom Johnson assured him that if some outside organization paid his salary he would be welcome to continue his research at the reactor.

He visited Bell Labs for a job interview. Shockley and Brattain greeted him and described their future Nobel Prize work on the semiconductor silicon.

"What can you do with neutrons to help us develop silicon?" asked Shockley.

Weiss didn't know and Shockley rephrased the questions, "What can you do with neutrons?"

Again Weiss shook his head, too much of a greenhorn to make any profound statement other than to briefly describe his thesis. GE, Sylvania, and probably a few others were approached but failed to take the bait and time began running out for Weiss. Tom Johnson came to the rescue and contacted Dr. H.H. Lester, an old friend at the Watertown Arsenal, an Army Ordnance research and development center near Boston. Lester's MIT-educated group leader Dr. Leonard Jaffe was willing to take a gamble and Weiss signed on as a civil servant and was given orders to continue his neutron research at Brookhaven. While working for the government was considered demeaning for a research physicist the terms of employment were sufficiently advantageous for Weiss to swallow his pride.

The research aura that permeated science in those days made few demands on the scientist to produce anything socially redeeming. Weiss found himself being supported to fulfill the whims of his own imagination, responsible partially to his immediate supervisor but principally to some higher authority, the laws of nature. For twenty years this atmosphere continued in America until it was finally realized that research budgets would someday outstrip the GNP. Today the 'bottom line' profit motive plays a dominant role, causing the laws of nature to be 'stretched' or 'bent' to sell one's research.

The halls of the physics building at Brookhaven became the venue for considerable chitchat as scientists passed each other going to or from the loo. Sam Goudsmit would always be willing to stop and gab. Sam had found a shop that had some remaindered copies of his book ALSOS. He bought the lot and sold them autographed at cost plus one cent to those interested. The one-cent was the charge for his signature. Weiss avidly absorbed its contents as it described the search for the German 'Oak Ridge', the roundup of the ten leading German nuclear physicists including Heisenberg, Hahn, von Weizsäcker, von Laue, Gerlach, et al., and the internment of the group at 'Farm Hall' near Cambridge. In both Goudsmit's book and his private chats with Weiss he frequently alluded to remarks the Germans made during their imprisonment.

"How did you know what they said?" asked Weiss.

Goudsmit merely smiled.

Day after day Weiss pushed Goudsmit for an answer and finally received one.

"The place was bugged," whispered Goudsmit.

"Wasn't that illegal?"

Goudsmit again smiled. In spite of the contravention of the Geneva Convention rules R.V. Jones, Churchill's fair-haired young scientific authority, arranged for those listening devices. For almost fifty years the transcription of those wire recordings were kept secret from the world by the British Foreign Office. In later years Weiss had discovered that Jones shared a fascination for Benjamin Franklin and they met for lunch. Alas! Jones would say nothing about the bugging incident.

During the immediate postwar period an emotional public dispute arose after the German scientists were released from their detention at Farm Hall. Heisenberg was quoted in the press as denying that the Uranium group he and Gerlach headed were aiming toward bomb development, only peaceful uses. Goudsmit had lost his parents in a concentration camp and was emotionally repelled by Heisenberg's profession of innocence. In addition he had been privy to the secret recordings although unable to admit to their existence. Before Weiss left Brookhaven he attended a talk given by the visiting Heisenberg with an introduction by Goudsmit. Sam indicated that the two had had their differences but had decided to bury the hatchet. Weiss eventually wrote a radio dramatization about the Farm Hall story which the BBC performed a year before the transcriptions of the recordings were released in 1992. The subject is still controversial and is told in great detail in the book *HEISENBERG'S WAR* by Tom Powers, Knopf, New York, 1993.

During this period Goudsmit took over the editorship of the Physical Review with Si Pasternak, his assistant. On one occasion Goudsmit had received an enormously long theoretical paper from John Wheeler on the theory of fission. Goudsmit remarked that if it takes 100 pages of equations in the Physical Review to explain fission it is clear that Wheeler didn't understand it. That has a lovely ring of truth like the remark that 'mathematics is a contraceptive to understanding'.

Goudsmit once remarked that he had a photographic memory and Weiss called him on it. Drawing a complicated and random maze of lines on the blackboard Goudsmit studied it for a minute and Weiss covered the sketch. A month later Goudsmit was able to do an impressive job in reproducing the figure.

Shortly after Goudsmit assumed the editorship of the Physical Review a paper was submitted by several professors in the Mathematics Department of one of the Associated Universities. They had performed an ESP experiment on guessing the turn of a card and had analyzed the data according to accepted statistical analysis. The editor could find nothing wrong — they had shown a small but positive effect. What to do? Should they publish? The editors huddled together but chickened out. If they gave credence to the effect it would imply that scientists could influence the outcome of an experiment by wishful thinking (psychokinesis). That is a no-no in physics. Only God can make a tree.

Lawn parties were popular at Brookhaven and permitted Weiss the opportunity to converse with the famous like the soft-spoken Eugene Wigner, the flamboyant Abraham Pais, the mesmerizing Robert Oppenheimer, the imperious Edward Teller, the Lincolnesque Glenn Seaborg, the Napoleonesque I. Rabi and many more. What an introduction for a young physicist to the celebrities of the day!

An almost legendary MIT scientist, John C. Slater, arrived at Brookhaven to spend a year and was provided an office next to Weiss. He brought four of his students with him, Kleiner, Parmenter, Schweinler, and Koster. When Weiss asked why Slater chose students whose names all ended in ER Slater confessed that he had never taken notice of that detail. Weiss estimated the random probability for such an occurrence as greater than a million to one.

Slater presented a series of lectures about theoretical atomic physics, delivered in one of the most flawless styles Weiss had heard. In Weiss' opinion only the recent presidential science advisor, D. Allen Bromley, has ranked with Slater in smoothness and clarity of delivery.

This proficiency in the command of English also extended to Slater's writings. He has been known to type out many of his textbooks in a day or two. Weiss later learned that Slater had arranged the Brookhaven visit to establish residence requirements for his divorce. No matter, the subject of Slater's tutorials made Weiss an aficionado for the subject of atomic structure and he looked forward to leaving Brookhaven in order to return to Watertown and perform x-ray determinations of electron positions on atoms. If Weiss had any premonition as to how much trouble the work would entail he might have tried something else. In any event he had had his fill of neutrons and was anxious to try something new.

Quantum mechanics makes several predictions about electrons on atoms, principally their energies, their positions, and their momenta. Much of the early success of quantum mechanics came with its ability to calculate the quantized energy levels associated with the optical spectral lines emitted when atoms are excited. The positions the electrons pass through as they whirl about atoms is also predictable but no one had made an effort to measure this by scattering x-rays from atoms. When Weiss informed Slater that he was determined to take a shot at this measurement, he was given much encouragement. An experimentalist might have talked him out of it. Fools rush in …

Weiss' neighbor at Brookhaven was a young physicist Arthur Vash. They met often and developed a friendship. When Weiss left Brookhaven the two lost all contact for <u>forty years</u>. Standing with an acquaintance in a crowded Back Bay RR station in 1990 Weiss had occasion to laugh and was soon approached by a 'total stranger'.

"I came over because I recognize the laugh. Are you Dick Weiss?"

"Yes."

"I'm Arthur Vash."

Having lost all his hair Weiss could not recognize Vash nor could Vash recognize Weiss. But to identify a laugh after four decades amazed them both and underscored some interesting physics about the auditory storage patterns in homo sapiens. Humans can recognize hundreds of voices and thousand of faces in less than a second. How is the information stored and retrieved? Scientists really don't know.

During Weiss' Brookhaven sojourn he was telephoned by one of the senior scientists at Watertown on June 29, 1950 — the fiscal year ending on June 30.

"Dick, this is Larry Foster. I just got a call from Washington telling us they have a quarter of a million dollars left over in the budget. Can we spend it in one lump sum by tomorrow so we won't lose it?"

"Interesting! Do you want <u>me</u> to spend it?"

"No, but I was thinking of getting a van de Graaff accelerator — 2 million volts."

Weiss mused over this, "That's almost ten volts per dollar. What are you going to do with it?"

"I don't know but we could hire someone to use it …"

Weiss grinned, "If it has to be spent in one chunk and it's a sin to lose it, I have no better suggestion."

Larry hung up and called Weiss two days later.

"The lawyers and I sat down with High Voltage Engineering and we signed a contract for an accelerator. I asked for one with two heads so we can either accelerate protons or electrons. They never made one like this before but they signed the contract anyway."

"Very good," said Weiss.

Larry hung up and called Weiss a month later.

"High Voltage Engineering tried to design an accelerator with two heads but concluded it was too difficult. They decided to give us two separate accelerators for the same price — one for electrons, one for protons."

A few months later High Voltage Engineering delivered the two machines but the purchasing officer at Watertown refused to accept them. He insisted that the government contract called for one machine and that was that!

High Voltage Engineering sent their delivery van to pick up the two huge machines, returned them to the factory, bolted the two control panels together, rewrote the bill of lading and made a second delivery. This was accepted as a single machine by the purchasing agent. The story still had a few twists. Alas! There was no money allocated for shielding the two monsters against stray radiation emitted when the machines were energized and the machines sat in the Watertown warehouse for years. One of the machines was eventually given to Stanford Research Institute. Sic semper tyrannus!

During this period Cliff Shull discovered antiferromagnetism in MnO, an ordered arrangement of magnetism on each Mn atom but with the north and south poles alternating in direction so that the substance could not be picked up with a bar magnet. But the neutrons were small enough to probe into the atomic structure of each manganese atom and to identify its magnetic direction. With this discovery a myriad of magnetic materials were soon uncovered through neutron diffraction and this field of study has continued to this day. Not only had Weiss been lured into measuring the electron positions on atoms with x rays but the position of electrons in magnetic materials became even more fascinating.

It had never been understood why ferromagnetism was limited to certain elements like iron, nickel, cobalt, and gadolinium — a mere 4 out of 100 or so elements. Theoreticians like Slater and Heisenberg had advanced some ideas to account for this and the scientific community accepted their explanations. But once neutrons began probing magnetic substances it was soon realized that the subject was infinitely more complicated than predicted by theoreticians. Einstein had said that, "Things should be made as simple as possible, but no simpler." It is becoming clear that magnetism like superconductivity is more subtle than imagined, and continued probing only raises more questions than it answers. Even though both the theories of superconductivity and magnetism have received Nobel Prizes, the Nobel committee commands no special pipeline to the Deity and the awards are not revocable.

Living next door to Weiss at Brookhaven was a graduate student from Hartford, Walter Knight. One evening he arrived home all excited since his magnetic resonance measurement had detected copper in its metallic form. It was believed you couldn't make a measurement on a metal since they conducted electricity too well. Surprise, surprise — if you looked carefully you could clearly find it although the magnetic resonance was shifted. It came to be known as the Knight shift. Just as Serber had insisted Fowles and Weiss were barking up the wrong tree, so Knight had been influenced by an expert who had failed to realize that nature is always full of surprises. Beware of authority! An expert is nothing more than a son-of-a-bitch from out of town!

Two experiments during this period stand out in Weiss' memory. Soft-spoken Andy McReynolds measured the gravitational free fall of the neutron, not that he expected anything other than g but who knows? An answer other than g would have become a Nobel Prize candidate. Employing a well-collimated beam and a long mean free path that required him to place his detector in the next building, he measured the position of a beam of fast neutrons that would fall very little in the earth's gravitational field and compared this with a beam of slow neutrons, the latter taking considerably longer to reach the detector. In the end McReynolds claimed that his determination of g from the neutron's free fall was certainly the most inaccurate measurement of that quantity.

Don Hughes and Harry Palevsky devised a clever technique to measure the scattering of the (uncharged) neutron by the (negatively) charged electron. No one knew whether such an effect existed although Fermi had postulated a model for the neutron in which it was sometimes a proton and a meson. The proton would, of course, be scattered by the electron. By employing a bismuth mirror in both air and immersed in liquid oxygen the critical angle for total internal reflection was measured. The 83 electrons on the bismuth atom yielded a significant contribution to the critical angle. About this time L. Foldy pointed out that the moving dipole moment of the neutron would cause a predictable scattering from the electron. This turned out to be the size of the measured effect leaving naught for the Fermi postulate, a 'mystery' to this day.

During Weiss' last year at Brookhaven the wrath of Senator McCarthy blew through science like the Inquisition. The Oppenheimer-Teller conflict and the fall from grace of J.R. Oppenheimer assumed crisis proportions for Weiss. He had lived through the Depression, shielded by his parents, but this 'red scare' loomed larger in severity. McCarthy had induced a polarization in our society that turned friend against friend. People who had innocently cavorted with left wing ideas in the thirties were losing their security clearance and their jobs. Weiss was fortunate. He had been much too engrossed in playing baseball during his college days to have signed any left wing petitions. Luck was on his side.

The American Physical Society is the prime organization for arranging meetings and publishing research papers for its members. Following the war their numbers jumped precipitously to reflect the national reverence for a group of physicists that had unleashed its own messiah. It became fashionable to attend the meetings of the Physical Society, deliver ten-minute research reports, and to read the physics journals. Washington D.C. was one of the more popular venues for these meetings.

Weiss met a naval officer in one of the corridors of the Wardman Park Hotel and asked him whether any work was being considered to construct nuclear powered submarines, a natural question for one who had been a marine engineer and was now involved with reactors. The advantage in sustained underwater operation without the need of oxygen for combustion was obvious. But this subject was hush-hush at the time and Weiss was quickly ushered along the corridor and introduced to a Navy Captain with the odd name Rickover. After an introduction Rickover unmercifully lashed out at Weiss.

"Why aren't you physicists working on nuclear powered submarines?"

Navy captains were not Weiss' favorite breed of person but he still respected their authority, although preferably at a distance. Rickover's caustic dressing down continued and Weiss was at a loss as to what it was all about. He eventually backed away from Rickover. At the time the Captain was beset with the concept of a nuclear submarine but failed to realize that one hardly ingratiated oneself to an academic by first imposing his authority. Rickover pursued his plans with great zeal but even other Naval officers found him abrasive. He eventually was promoted to admiral after being passed over three times but only because congress interceded. While Rickover is given credit for the success of the nuclear submarine program Weiss felt that a more genteel approach might have worked to better advantage. Actually, physicists were fed up with military work and wanted to get back to pure research. Rickover was frustrated at the lack of interest in the project and approached his conflict with physicists as a military engagement.

Weiss recalls another American Physical Society meeting at this time, this one turning into a disaster. He had journeyed to Columbus, Ohio, a town that boasted the campus for Ohio State University on its outskirts and a grotesque prison in the center of town. The Deshler-Walleck Hotel played host to the Society but the management failed to realize that physicists were a different breed of guest. Their mistake came with their standard over-booking of rooms by 20% to allow for no-shows. All the physicists turned up for the meeting. What a panic finding rooms elsewhere, even renting the suite otherwise reserved for Senator Robert Taft in the hope the Senator would not appear in town.

Weiss was queried by the taxi driver taking him to the airport after the meeting,

"What's wrong with these guys?"

"What do you mean?"

"What do these guys sell?"

"They're physicists — they make atomic bombs."

"Holy shit!"

"What's wrong?" asked Weiss.

"For one thing, the bartenders have been complaining that they don't drink — and they don't tip, either."

"What's the second thing?"

"They haven't used any of the girls."

Weiss smiled. This was a thumbnail sketch of a community of physicists; virtually all male, with little money to splurge on drinks and tips, and too shy or poverty stricken to engage in horseplay. They were the model of pristine virtue. They played tennis but not golf or baseball. They listened to classical music and played musical instruments, particularly enamored of the mathematical precision of Bach. They did not engage in play-writing or acting, both being too emotional. To many they even fit the nerd description. Admire them, if you must, but from afar.

WATERTOWN 1: 1953–1956

Richard moved to Watertown with a delightful feeling of anticipation and of returning to civilization. The arsenal was a short drive to both MIT and Harvard and was situated on the Charles River but a few miles upstream from the cultural center of America-Boston. As one grand dame of the Brahmin establishment who had never left her home on Beacon Hill put it,

"Why travel when I'm there already?"

The national enthusiasm for physics research in the 1950's was engendered by the congressional and military awe over the physics geniuses that produced the atomic bomb, further fueled by the mushrooming cold war. The Capitol lawmakers provided the US Army with an adequate budget for feeding the cranial mills of scientific thought as well as funding an overkill of military hardware. Richard's colleagues at the universities wondered how guns and armor could be improved by a bunch of physicists measuring electron distributions but spared him the embarrassment of posing this question. The Generals at the Pentagon didn't know the answer (neither did Richard) but it was fashionable to support such basic research and none of the Army brass wanted to appear out of step with congressional and national attitudes. Nary a soul in the chain of command from the Director at Watertown Arsenal to the General Staff at the head of Army Ordnance in the Pentagon ever asked during the 1950's how the knowledge of electron distributions, Richard's avowed research intention that he promised Slater, would make a better gun or a more resistant armor plate. But if these longhairs could make an atomic bomb, why not a better gun? The day Richard arrived at Watertown he passed the Commanding Officer who shook his hands, grinned, and quipped,

"We've all been waiting for you."

The inference was clear — Watertown now had its very own physicist to parade before the powers to be. Now that he had arrived things were going to improve.

Richard occasionally spewed some nonsense in informal discussions with the brass to the effect that better knowledge of electron positions would result in designing better metals but in his heart he knew it was a delusion. It was certainly what they wanted to hear but too timid to ask. Richard was reminded of the group of physicists in Germany seeking funding during the war for a centrifuge that could only separate micrograms of isotopes. They justified the work by claiming that the lighter isotopes of iron would reduce the weight of tanks! It was sheer mendacity that any freshman physics student could see through.

In dealing with the military, the Watertown physicists acted disingenuously, smug in the belief that the generals didn't know their derrieres from les coudes, and even having a private laugh over it. It eventually caught up with them 20 years later when congressmen began asking pointed questions. One Senator was advised that physicists needed a place to sit and think.

"It's easy to tell when they're sitting but how do you know they're thinking?" countered the solon who wanted to know what he was paying for.

The Watertown Arsenal of 1952 was a typically bland example of Victorian arsenal architecture comparable in moroseness to the mills of Lowell and Fall River. It dated back to 1816 when the Army needed an accessible New England location for the storage of arms. The chosen spot, where the Charles narrowed at a bend in the river, had been a site for Indian settlements for thousands of years as revealed in archeological digs. With each war and the ever-increasing demands for armor and guns, more buildings, land, and facilities were added to the original Watertown site. By the time of the Civil War the arsenal staff had increased to almost a thousand. The importance of metallurgical inspection and improved steel manufacture was underscored with the occasional tragedy of flawed gun barrels blowing up and killing the gun crews. Just after WWI Watertown

Arsenal had improved x-ray radiography and other testing technologies, eventually excelling other establishments in the quality of its guns.

Watertown could fabricate gun barrels by centrifugal casting, a process that required less machining since the hole was partially bored during solidification. The Arsenal also funded the many post WWII university research programs that developed titanium alloys as a viable structural material. These were lighter than steels and could withstand higher in-service temperatures. During WWII the demand for the Arsenal's products was so great that the facility expanded its work force to over 10,000 and increased its acreage to 130. Even President Roosevelt and Douglas MacArthur acknowledged Watertown's importance with personal visits.

Within a few years of Richard's arrival at Watertown students of both John Slater and the well known MIT x-ray experimentalist Bert Warren sought employment there since it would enable them to continue to do interesting physics and to remain in the Boston area after receiving their PhD's. Art Freeman, Art Paskin, Lee Allen, Tony Hofmann, Dave Chipman, Chris Walker, Homer and Grace Priest, Ken Tauer, Ken Moon, Paul Sagalyn, and Bill Croft were taken on to create an 'elite' group of physicists. Indeed, while many other scientists joined Watertown during this period the basic research group was eventually given an exclusive use of its own building as well as the authority to keep out newcomers that didn't meet their perceived standards. This naturally created antagonism between this 'private club' of cogniscenti spewing unintelligible physics and other scientists pursuing more mundane endeavors like developing better materials for the Army. Richard cannot cite a single instance over a 20-year period where this group of savants did anything useful for the Army except garnering a little prestige in the world of basic research.

Leonard Jaffe, a metallurgist of some ability, served as the immediate supervisor for this aristocracy of scientists but he bore the tiring task of having to report to a supervisor who delighted in providing Washington with innumerable reports and in running a tight ship. The Supervisor's knowledge of science was miniscule and Jaffe did his best to shield the group from this 'paper tiger'. When the group moved into a newly renovated building, formerly the Simmons mattress factory, Jaffe had opted for a corner office with a built-in intercom system that enabled him to communicate with the offices and labs of those within his group. Alas, someone with 'pull' at high places took a shine to Jaffe's office and he was evicted. In order to maintain the intercom network, Jaffe personally took several days to rewire the system, leaving cables exposed and hanging from the ceiling. On the day the Generals arrived from Washington to inspect the new building the 'paper tiger' lowered all venetian blinds to half mast, turned on all fluorescent lights whether the office was occupied or not, opened all doors fully, asked everyone to polish their shoes, had the floors polished and the walls cleaned, and removed Jaffe's exposed wires. Jaffe became so distraught he took a pair of shears, entered the 'paper tiger's' office and severed his intercom wires — 'an eye for an eye'.

At the official opening reception of the new building one of the second lieutenants on the post brought in a turntable to play the national anthem, a live band not being available. After the pledge of allegiance the record player was started but out blared *GOD SAVE THE QUEEN*, familiar but not correct. Following a rapid shuffle the record was turned over to play the *STAR SPANGLED BANNER*. What a gaffe!

A short time after his imbroglio with the 'paper tiger' Jaffe announced his resignation to take up a position on the west coast. Richard asked Jaffe to also find him a job but the California based Jet Propulsion Lab could only offer him a research position without a raise. Richard turned it down since living costs were obviously higher in Pasadena. This move would have placed Richard in the space program with no chance to fulfill his promise to Slater.

The gradual expansion of the physics group and the proliferation of its publications in the respected journals like

Physical Review reduced the indignity associated with being part of a government lab. On one occasion, though, Arthur von Hippel, a well-known professor of material science at MIT, confided to Richard his belief that government laboratories should get out of basic research and turn the activity over to the universities. On first hearing this, Richard was taken aback since he considered the elite group at Watertown comparable in research output to similar groups at universities. In retrospect Richard believes that von Hippel may have been right but in those glorious days in the 1950's Richard and his colleagues enjoyed their physics research too much and worried but little as to whether they were doing their bit for the Army during the cold war. Furthermore, the Army brass continued to boast about their team of physicists.

But it was the very nature of the Civil Service that was responsible for its disorganization since Washington was staffed with scientists who couldn't perform well amidst the technical complexities of their own subject, compelling them to migrate to administrative jobs in D.C. Unfortunately, these people called the shots when it came to funding. Richard thought he had found a compromise when the well known Cal Tech metallurgist Pol Duwez, who had worked with Army Ordnance during the war, agreed to come to Watertown and head up the basic research activities. Unfortunately the same individual who had relieved Jaffe of his corner office managed to squelch this appointment. Except for the Naval Research Lab in Washington Richard found government labs to be a haven for the less talented. Through the tenure system scientists found a 'home' and it was impossible to weed out the incompetents. Forty years passed and nothing changed.

Actually, to salve their consciences for not contributing to the Army's needs Richard and several of his colleagues asked the Director at Watertown to identify some Army problems to which the physics group could apply their talents. A few weeks later the Director suggested to the group that they find a use for the element silicon.

"It was cheap, plentiful, and light weight," he said.

Richard considered this a surprisingly worthwhile suggestion emanating from the Director. The group withdrew to contemplate their navels but they concluded that due to its brittleness silicon had no practical use in Army applications. If you couldn't make a gun out of it what could you do with it?

Richard reported back to the Director,

"We think God made a mistake with silicon."

Prior to the discovery of the semiconductor, silicon was a bit of an unwanted child in the scheme of things. If Richard's future advice had remained as consistently wrong as he was about silicon he would have been invaluable to the Director and the Army! Richard should have remembered that when he had applied for a job at Bell Labs in 1950 the future Nobel Laureates Shockley and Brattain had shown him their work on silicon P-N junctions but he was too much of a greenhorn to understand its significance. That was the only time the Director challenged the group.

After settling in at Watertown Richard and the hard-working John DeMarco set out to examine the arrangement of electrons in iron by employing x-ray diffraction. They chose iron because it was interesting in itself and no General would query the choice since it was the main ingredient in steel. But there was virtually nothing in the scientific literature that provided clues to the difficulties that might be encountered in employing x-rays to answer that question. Research invariably depends on the experiences of those who went before. With very little to go on and with no one offering specific advice about the pitfalls, Richard ordered the latest x-ray equipment and prepared some iron samples for his trek into the unknown.

The theoretical idea was simple enough. If x-rays of a single energy are scattered by isolated atoms the intensity versus the scattering angle depends in an elementary way on how far from the nucleus the electrons are distributed, the very property people like Slater's students at MIT and Hartree at Cavendish Lab in England had tried to calculate. But iron atoms practically always appear in solid pieces. (You could try to vaporize some iron but the high temperatures needed made it difficult.) The iron atoms in a piece of metal were in a crystalline array and this introduced additional problems that unrelentingly plagued Richard. It no longer became a straightforward matter to relate the scattering of x-rays to the electron distribution.

But having started down this road and being driven by intellectual curiosity, supposedly the fundamental mark of a dedicated physicist, Richard spent five years seeking the answer to both the electron distribution and how this arrangement of electrons accounted for iron's ferromagnetism. As if this property of iron didn't prove difficult enough to understand there is a second metallic form of iron present at high temperatures that is not ferromagnetic, but one must add some manganese to stabilize it at room temperature. How the electron arrangements differed in the two forms of iron was a question of intense interest to those concerned with the subject called solid-state physics.

Serendipity came into play when Richard tried to fabricate samples of this non-magnetic iron to compare with the garden variety magnet form. He requested the Watertown foundry to melt together pure iron with the necessary amount of manganese and carbon but was informed it was a tricky procedure to get an alloy that wasn't a mixture of the two forms of iron. Actually, proper fabrication required cold rolling as had been demonstrated before WWI when the so-called Hadfield steel was discovered in England. It was shown by Hadfield that this non-magnetic form of iron, containing about 12% manganese and about 1% carbon, was completely non-magnetic and very tough, indeed. Try cutting it with a hacksaw and the blade was quickly rendered dull. By fabricating helmets out of this material Hadfield boasted that thousands of British lives were saved in WWI.

But before the reader comes to the conclusion that the strength of Hadfield steel is related to the (lack of) magnetism dispel the idea — it's not. Richard was anxious to obtain a sample of this non-magnetic iron and was advised by a metallurgist at

Watertown that there were plates of Hadfield steel (helmet metal) left over from WWII and they were sitting in a Watertown storehouse. He secured several sheets, cut out a few samples with a torch, and brought them down to Brookhaven to be examined with neutrons. Neutrons, copiously generated in an atomic reactor, are sensitive to the magnetic structure of atoms since the neutron behaves like a small magnet. What a surprise to discover that at liquid nitrogen temperature Hadfield steel was antiferromagnetic i.e. half the iron atoms were magnetic with their north poles pointing one way and the other half pointed in the opposite direction. The net effect is that they cancel each other and Hadfield steel is not attracted to a magnet such as ordinary iron.

This was one of those discoveries that delight scientists but it turned out to be only one in a long series of experiments that revealed iron to be a most complicated element. Strange that one of the most abundant and useful metals in nature should present the solid-state physicist with innumerable theoretical and experimental complications!

After years of relentless effort and weekly discussions with Slater and Warren, Weiss and DeMarco came to the conclusion that the electrons responsible for the magnetism in ordinary metallic iron were isolated from the other electrons. This unexpected result impacted the physics community with astonishment. There is an old saw in science that there is nothing so startling in physics as a mistake! Richard never felt comfortable with the result and took a few months off to visit several laboratories to present his findings and to discover if he had overlooked something. Zachariasen in Chicago and Pauling in California listened but remained non-committal — they really had no ideas.

After his return to Watertown Richard made it a habit to show up at MIT on Wednesday afternoons and both Warren and Slater would kindly drop what they were doing and exchange thoughts over the Watertown experiments. Within six months they began to believe the unusual results and this provided

Richard with some unjustified reassurance. Richard was so frequently seen with Warren that most people in the x-ray community assumed he had been one of Warren's graduate students.

But the more Richard delved into the electronic and magnetic structure of the first group of transition metals i.e. elements from titanium (atomic number 22) to nickel (atomic number 28) the more he came to realize the complexity of their electronic arrangements. Furthermore there was a strategic fault in these research efforts. Implicit in the pursuit of knowledge was the belief held by Richard that he would ultimately unlock the secrets of the electron structures (whatever those secrets were) and be able to tailor new alloys and compounds with predictable properties. Not so. Every time Richard drew back the curtain a soupcon he uncovered more problems. Answer one problem and create ten new ones. Research was like a chain reaction leading to a brilliant explosion of ignorance.

About 1955 Richard ran out of ideas for further experimental checks on the <u>unexpected</u> arrangement of the electrons in iron and opted for a change of pace. He wasn't ready to publish the work and he let the matter rest. But he remembered the Zener lecture at Harvard that had related the specific heat measurements of iron to its ferromagnetism. It was particularly intriguing since it provided an independent check that only 2 of the 26 electrons on each iron atom in the metal gave rise to its ferromagnetism, already known from iron's magnetic strength. The magnetic entropy, deduced from the specific heat, could be related to this number. Three others in the research team, Tauer, Paskin, and Hofmann, joined in the fun as the literature was searched for other substances that showed a similar pointed cusp in its specific heat.

In working on the problem Richard naturally developed a close rapport with the others in the group. In particular, he learned that Tony Hofmann was the son of the famous pianist Josef Hofmann, a child prodigy who was brought to this country in the 1880's to give his first concert at Carnegie Hall at age 8.

Tony, himself, never showed any talent for the instrument and was discouraged from pursuing it after his father heard him play at age 4. Funnily enough, a popular high quality loudspeaker sold under the name KLH contained the letter H since Tony Hofmann was one of the three inventors. It relied on the air in a sealed speaker to provide the restoring force as the diaphragm was driven by the coil (rather than any mechanical device providing the restoration).

Josef Hofmann created a sensation when he appeared in America as a child prodigy. His father was loudly criticized for exploiting and overworking the child although he claimed that playing the piano was fun to Josef. Some skeptics even suggested that the child was a midget! When asked to play a new piece of music that he had never seen before, Josef Hofmann's father responded that the child was not old enough to read music!

A solution was reached when a Pittsburgh steel magnate stepped forward and anonymously provided the father with funds to remove the child from the concert stage until age 21 and, until that time came, have him study under the famous Anton Rubinstein. (Tony was actually christened Anton in honor of the pianist but preferred the more popular Tony). Josef Hofmann returned to the concert stage at 21 never having learned the name of his patron. But when Anton Rubinstein died Josef Hofmann gave a memorial concert in Liverpool. One member of the audience was so impressed that he approached Josef Hofmann after the concert and revealed that he was, indeed, his patron and just happened to be in Liverpool at the time! It turned out that Josef Hofmann was also a successful inventor and probably passed on this talent to Tony in lieu of music.

After exhausting the literature and corroborating that the number of magnetic electrons in various magnetic substances could be determined from their specific heat Hofmann, Paskin, Tauer, and Weiss were ready to publish. At the end of 1955 a new journal on the chemistry and physics of solids was announced by Robert Maxwell at Pergamon Press with Harvey Brooks of Harvard as the editor-in-chief. With such a distinguished overseer Richard felt that Maxwell's questionable background was now behind him and the paper was proudly submitted for the very first issue of *THE INTERNATIONAL JOURNAL OF THE PHYSICS AND CHEMISTRY OF SOLIDS*. As nearly as he could remember Richard found that the published results had little impact on the scientific community. But this letdown comes to all, whether scientist, writer, or creative performer. You may shout your accomplishments from the rooftops but everyone's out of town.

This was food for thought. The average scientific research paper was referenced on average by less than one other research team who subsequently published in the field. Presumably learned journals existed to help those who follow on in the subject. If very few other scientists benefit from a publication one might question the logistics of filling libraries with these journals. They were costly, they occupied much space, and they were 'unreadable'. After all, thought Richard, when a university granted a PhD in physics it did not assure the world that the recipient could write or speak effectively. Yet, if the drama or English departments were called in to certify the candidates' ability to communicate effectively, it might force universities to take appropriate action in order to produce more eloquent PhDs. Unfortunately, departmental pride made such cooperation unlikely.

When the Physical Review editorial offices moved into the physics building at Brookhaven Richard often found himself called upon to referee papers. It was bad enough that the physics was complicated; almost always sprouting (and spouting) forbidding mathematical formulations, but the use of the King's English was abysmal. No wonder very few papers were read. Sometimes Richard was asked to 'translate' an English paper from Japan into acceptable prose. During this period Richard wrote a letter to the editor: Why not pay referees to do a thorough job of reading and editing and why not identify the referee? Chief Editor Sam Goudsmit wrote a stinging rebuke — referees must remain anonymous and no money must taint their tasks! Bang! Shot down right between the eyes.

The proliferation of published material reminded Richard of one professor at Penn State who was sent reprints of papers by his former students and other colleagues he had befriended over his long career. He wouldn't discard any of these and they eventually were stacked from floor to ceiling in piles that occupied all spare footage. To enter his office was to enter an arsonist's heaven (or hell if he succeeded). On one occasion this professor couldn't find his telephone since even his desk became a tower of reprint babel. What to do? He went into the secretary's office and asked her to ring him. Voila! (sorry, écoutez!)

Of the many organizations that gave scientists an excuse to visit a new locale and make new friends the International Union of Crystallography (IUCr) boasted a large membership. By announcing a postwar meeting in Paris in 1954, their third international convention, the busy bees of research were attracted to the honey. At that time overseas travel for government employees was not frowned upon since the Defense Department ran their own airline MATS, Military Air Transport Service. There were no frills, it was cheap, and the DC-6 aircraft that flew from Westover AFB in Massachusetts to Frankfurt were sufficiently uncomfortable to remind one that the inconvenience reflected the cost. Richard gladly opted to withstand the tiresome journey in order to see Paris and to rub elbows with scientists who had gained fame in the literature.

Of the 800 attendees a significant number had left their mark in crystallography and some had received Nobel Prizes. These included Bjelov of Russia, Bose of India, Brill from Rome, Wilson from Cardiff, Wyckoff then in London, Taylor from Cavendish Lab, Perutz from Cavendish, Wilkins from King's College London, Lowde from Harwell, Lonsdale from University of London, Hodgkin from Oxford, Rosalind Franklin from Birkbeck London, Sir Lawrence Bragg from London, Bernal from Birkbeck, Bacon from Harwell, de Broglie from Paris, Guinier from Paris, Friedel from Paris, Curien from Paris,

Warren from MIT, Slater from MIT, the Karles from Washington, Fankuchen from Brooklyn, Ewald from Brooklyn, Barrett from Chicago, and von Laue from Berlin. The large contingent from England not only made the journey to Paris because it was short but also to seek out a few good meals, England still rationed meat.

Prince Louis de Broglie had received the Nobel Prize in 1929 for postulating in 1923 that matter, as well as light, possessed a wavelength, thus providing the clue that led Schroedinger and Heisenberg to discover quantum mechanics. His opening speech, in French, reminded Richard of the conference he attended in Paris at which earphones were provided at the entrance, displaying the sign:
RECEPTEUR FOR TRANSLATION.

Richard advised the secretary of the proper English word for recepteur. She thanked him and the sign was then changed to read:
RECEIVER POUR TRADUCTION

To paraphrase a portion of de Broglie's invited talk at the Sorbonne:

I am embarrassed to give a talk before a gathering of specialists in crystallography, not having any knowledge of the subject. For that reason I will discuss the use of crystals to demonstrate the wave nature of light.

The first experiment by physicists revealing the symmetry of crystals was the discovery of double refraction in Iceland spar by Bartholin in 1669. At this time one didn't understand crystal structure until Huygens postulated the wave nature of light. But for a long time scientists didn't follow up on the propagation of light in crystals.

In the 19th century l'Isle, Haüy, and Bravais began to understand the structure of crystals from their symmetry. Later Young and Fresnel showed how the wave nature of light led to diffraction and in 1807

Malus postulated the existence of polarization. By treating light as a transverse vibration (transverse to its direction of propagation) Fresnel and Huygens developed the theory of light propagation in crystals that still exists today. The discovery of the rotation of the plane of polarization and the effects of chromatic polarization have provided insight into the internal structure of crystals. Both Pasteur and Pierre Curie have stressed the importance of symmetry in physics and biology.

It was the discovery of x-ray diffraction by von Laue, confirmed by the Braggs, which uniquely revealed both the wave nature of x-rays and the structure of crystals, as given by the classical formulas of Laue-Bragg. This clearly reveals the arrangement of atoms when their separation is the same order as the x-ray wavelength.

While de Broglie scarcely uttered anything new to the audience it was exciting to see and hear a living legend, particularly coming from a member of the French nobility. It was also clear to Richard that crystallographers had neglected making measurements on the electrons that served as the glue to hold together atoms in crystals. New ground would have to be broken.

Richard had ample time to wander the streets of the City of Light and take in the touristy sights, now cleaned up a decade after the city's liberation. Perhaps the most memorable IUCr event during that convention was the banquet at Versailles. Imagine 800 guests under one roof all being served hot dishes at precisely the same time, course by course with military precision. The large crew of waiters, at least 100 of them, must have been trained especially for this event.

H. Curien, an x-ray physicist from France, attended the conference. He was later elevated to a post equivalent to a State Secretary for Science with control of the purse strings. In that position he had to deal with research proposals submitted by physicists. Asked how his new job differed from doing research, he joked to Richard that when two scientists submitted proposals for identical projects, he neither averaged nor added their estimated budgets to arrive at a true figure but multiplied them. Physicists, he warned, were notoriously inept at accounting. Even Einstein claimed he was poor at mental arithmetic.

After the conference George Bacon from Harwell, the British Oak Ridge, wrote to Richard to say that he was coming to America to make the rounds of the laboratories. George was then engaged in neutron diffraction and both he and Shull at Oak Ridge had examined pure chromium metal in powder form since neither knew how to obtain single crystals. They had each observed that chromium was antiferromagnetic i.e. each atom was a magnet but the north-south directions reversed on a regular basis. Hence there was no net magnetization as in a ferromagnet.

Richard picked George up at his hotel and George confided that he was at the end of a month's tour and this was the first night he had slept in a stationary bed. To save money the travel office at Harwell had booked him on overnight sleepers from city to city! George was gifted with an unusual power of recall, particularly items like old addresses, old telephone numbers, etc.

When Fermi had first performed neutron diffraction from simple single crystals he had noted that the intensity was not what he had expected in comparing the various orders. This remained a mystery until George Bacon and his colleague Ray Lowde made some extensive calculations that revealed the culprit as multiple scattering within the crystal. The effect became known as extinction and was the bane of all those doing neutron, x-ray, and electron diffraction. It was this very effect that was the undoing for Richard in the measurements he had made on iron. It varied from iron crystal to iron crystal and was virtually impossible to predict or measure with accuracy. Extinction is still a major problem today — that was the warning someone should have given Richard before he started on his x-ray measurements of iron but nobody but the Deity knew it and He wasn't revealing this secret to Richard without having him suffer.

At that time someone from the US Bureau of Mines sent Richard a single crystal of chromium and he took a few weeks off to go to Brookhaven to examine it at the Corliss-Hasting neutron spectrometer. The single crystal provided more scattered intensity than powders and enabled one to improve the measurements of the weak antiferromagnetism reported by Shull and Bacon. The particular peak (100) had an odd shape, much broader than the other peaks. At a suggestion of Hasting a plutonium filter was employed to clean up the neutron beam, which had a certain percentage of higher contamination. Voila! The (100) peak was actually two peaks split down the middle. This meant that the magnetic atoms were pointing in a more complicated pattern than just alternating north-south, they might even be arranged in a spiral. This surprising result was taken up be several theoreticians, including the future Nobel Laureate de Gennes who was visiting from France at the time, and a partial explanation was forthcoming.

As if this weren't enough to chew on, the specific heat of chromium did not obey the cusp-like findings outlined in the work of Hofmann, Paskin, Tauer, and Weiss. In fact, this problem and myriads of others about the 3d elements remained unresolved.

Richard enjoyed his trip to Paris, exploring its flea market, the old city, the small restaurants, the crisp bread, and the well-known sights. Following his return to Massachusetts two things occurred that promised to change his life at Watertown. Jack Goldman, physics professor at Carnegie Tech who was interested in the properties of the magnetic elements, was asked to start a research team at Ford Motors in Dearborn. He asked Richard to come along and a visit to Dearborn ensued. The major lure was the prospect of a reactor at Ann Arbor (University of Michigan) and the continuation of his work in neutron and x-ray diffraction. The offer was, indeed, tempting. But an announcement of the Rockefeller Public Service Awards enabling Civil Servants to spend a year abroad, caught Richard's fancy.

John D. Rockefeller III was the grandson of the famous oilionaire and he took up benevolence seriously. His grandfather amassed a huge fortune through his oil refining empire, ruthlessly eliminating competitors. In spite of this shady treatment of those who stood in his way he took up philanthropy with the huge fortune he had amassed, second only to J.P. Morgan. The appetite for giving away money continued with John D. Rockefeller, Jr. who helped establish the colonial village at Williamsburg as well as Rockefeller Center. With John D. III it was natural that he further the tradition. Thus one of his ideas was to reward outstanding government employees and he did this through the Rockefeller Public Service Awards.

Richard applied for the Rockefeller Award citing Slater as one of his references. And so it came to pass that on January 21, 1956 Richard received a letter from Harold Dodds the President of Princeton University, a post once occupied by Woodrow Wilson, informing him that he was one of 16 recipients of the Rockefeller Awards. A letter Richard then wrote to friend George Bacon at Harwell was passed on to the head of the metallurgy group who responded favorably at having Richard join them for a year.

Letters of congratulations were received including one from General Maxwell Taylor and many newspapers picked up the story. With the availability of Military Transport Richard flew off to England to see the people at Harwell about his forthcoming stay. He was shown about the laboratory and the nearby villages where he might find accommodations and was promised that the Housing Office would locate a place before he arrived. He returned home convinced that he would enjoy working at Harwell. He wrote to Jack Goldman turning down the offer to join him at Ford and set about making arrangements for the fall journey.

A few weeks later came a surprising letter of apology. As a matter of routine the invitation to have Richard spend a year at Harwell (at no salary since John D. Rockefeller's Foundation was footing the bill) was passed on to the Director at Harwell who turned it down,

"We don't want any Americans here!" he snapped or words to that effect.

What caused the bug to be nipping at the Director's derriere was never revealed to Richard but he immediately wrote to W.H. Taylor at Cavendish Lab in Cambridge to explain his plight. By return air mail he was assured that there would be no difficulty in his joining the x-ray group at the Cavendish. And that settled that! A week later the Director at Harwell who had turned Richard down died!

A follow-up letter from Princeton invited Richard to Washington for a banquet to honor the recipients of the Rockefeller Award. John D. III was present to congratulate the 16 awardees most of whom worked in Washington. After the ceremony Richard found himself on the same flight to Boston as the philanthropist and he asked him how he had come up with the idea for rewarding civil servants. Rockefeller pointed out that philanthropy was not easy. It took considerable thought to give away money so that it would reap a return. He knew many dedicated workers in Government Service and thought the reward idea worth trying for ten years or so.

CAVENDISH LAB, CAMBRIDGE:
1956–1957

A few years prior to Richard's arrival at Cavendish Lab, Francis Crick and a visiting American James Watson had teamed with Maurice Wilkins at King's College London and postulated the helical DNA structure. This was based on the x-ray data the BBC hinted might have been 'unethically' obtained from Rosalind Franklin in London on single crystal samples she had grown. Whispers of potential Nobel Prizes made the rounds in Cambridge but Richard was not a crystallographer and the biology of DNA was hopelessly beyond his comprehension. Complicated arrangements of atoms that were the milieu of the crystallographer left Richard bored — he was only concerned with the electrons that provided the 'glue' to keep atoms stuck to each other. He was definitely not drifting with the tide created by the biophysicists at Cambridge. When he was advised that their goal was to solve the structure of proteins Richard considered them foolhardy. He pressed ahead to tell anyone who would listen about his work on iron but it would take some months before his results would attract the enthusiastic attention of the Cavendish Professor and future Nobelist theoretician Nevill Mott.

Prior to the war such Nobel Laureates as J.J. Thomson, discoverer of the electron, Ernest Rutherford, discoverer of the nucleus, Sir Lawrence Bragg, discoverer of x-ray techniques to determine crystal structure, and James Chadwick, discoverer of the neutron, had trod the narrow Free School Lane on their daily walk to and from the famous laboratory in the heart of Cambridge. To Richard the town reeked with atmosphere, both scientific and literary. Newton (who discovered the gravity of things in the fall) and the poets Milton, Wordsworth, Coleridge, Lord Byron, and Lord Tennyson were all products of Cambridge.

As Thomas Gray had satirically written in 1736:

"Surely it is of this place, now Cambridge, but formerly known by the name Babylon, that the prophet spoke when he said, 'wild beasts of the desert shall dwell there, and their houses shall be of doleful creatures, and owls shall build there, and satyrs shall dance there; their forts and towers shall be a den forever, a joy of wild asses'."

The colleges had been in existence for ¾ of a millennium and the original Gothic architecture and manicured grass lawns attested to its age. The first college, Peterhouse, was founded in 1284. John Harvard had graduated from Emmanuel College before heading to Cambridge, Massachusetts where his gift of books became the foundation for the famous university on the Charles named after him. John Singleton Copley emerged from Trinity, Braddock, the General that Washington served under, came from Corpus Christi, and Roger Williams from Pembroke. There were over 13 colleges at Cambridge all somewhat self sufficient with a Vice Chancellor to rule on University matters and keep the colleges from drifting down the river. King's College Chapel, a gift of King Henry IV in 1446 but not finished until Henry VII's reign in 1515, was the architectural piece de resistance. It was almost unreal to the lad from the Bronx.

An American tourist asked the groundskeeper at one of the colleges how they achieved such beautiful lawns,

"You've got to use a roller every day — and don't miss a day for 700 years!"

The moderate temperatures and the frequent mists, sometimes called rain, were also responsible for the greening of Cambridgeshire. Richard shall always be indebted to John D. III and the Cavendish Lab for the opportunity to work in that town of high cultural density. The Bronx reared him but Cambridge enlightened him.

According to the terms of his Rockefeller Award he would be in Europe to study electron distributions — he barely got past the first fish and chips shop and the London Times the fish was wrapped in. Every day he met adventure in the buses, the trains, the Underground, the pubs, the newspapers, BBC, at afternoon tea, the unarmed bobbies, the language, the archaic money, the food, the telephones, the Royal Family, class consciousness, the dress code, etc. Until Richard became familiar with all these differences — well, you could get Richard out of the Bronx but could you get the Bronx out of Richard in less than a year?

Cambridge sat on the river Cam, the site receiving its name from the bridge crossing the narrow stream. Its paths were crowded with students seeking knowledge like an ant colony sniffing out sugar. Gowns were compulsory attire and the streets were patrolled by staff members to enforce this regulation. Occasionally an improperly dressed student was chased by the 'bulls' to nab and identify him. Eight-foot-long college wool scarves, each with its own identifying colors, were draped around the neck, oft times providing the prime shield against the biting chill. The absence of hills in the town, the narrowness of most streets, and the high cost of petrol gave the bicycle free rein. Even Professional window cleaners could be seen on bicycle with a twelve-foot ladder on one shoulder and a bucket and squeegee hanging from either end of the ladder. Dining at college often demanded dress attire for the dons, yet they would still rely on the bicycle for transportation.

Students and dons were separated at meals, the latter sitting on an elevated stage called the High Table (the food was superior, too). The dining halls were generally Gothic with high vaulted ceilings and a display of paintings that identified patrons and alumni. W.H. Taylor kindly sponsored Richard for membership so that he could share the meals and conversation of High Table at King's College Cambridge (not to be confused with King's College London which only dates to 1830). Someone pointed out the famous author E.M. Forster (*A Passage to India*) who was sharing the meal with Richard. It was almost beyond belief.

The market square in Cambridge accommodated about 100 stalls, a reminder of Richard's early years in the Bronx when his mother would take him to Bathgate Avenue and they would walk past the pushcarts that lined the street, trying to save a penny, half of which was probably expended in shoe leather. Vegetables, fruit, flowers, and doodads were dispensed to residents of Cambridge. Forty years later the proliferation of flea markets in America reminded Richard of the outdoor market stalls of Europe.

The presence of this Bronx-American at Cavendish Lab was unhappily noted the day he arrived as he was the only one who could afford to smoke cigars. He left a trail of pungent smoke and ash. The gossip soon made the rounds after Richard had seated himself in the austere Philosophical Library and broke the cardinal rule by lighting up a cigar. He was dumbfounded when he was asked to leave!

But without an x-ray machine at his disposal Richard wandered the streets, devoting but a few hours a day in the labs to theoretical problems in thermodynamics. He rubbed elbows with Brian Pippard who took an interest in Richard's work on thermodynamics, nodded to Otto Frisch who had first postulated the fission process with his aunt Lise Meitner in 1938, and he frequently consulted with Mott about the magnetic electrons on iron in attempting to arouse his interest.

Having already unearthed the possibility of reconciling magnetic structure with specific heat in the lengthy paper with Hofmann, Paskin, and Tauer, Richard had teamed up with Ken Tauer, his neighbor in Massachusetts and occasional golf partner, to decipher details of the two forms of iron metal. This was continued by post after Richard arrived in Cambridge. Unfortunately his former CCNY course in thermodynamics delivered by Zemansky had not aided him in a practical approach to the subject. It took Clarence Zener to lift the veil when he reported that the magnetic entropy of iron could be related to the number of 3d electrons, i.e. two per atom. At the time Zener awakened Richard's interest he had been at the Westinghouse research labs in Pittsburgh. But during WWII he had resolved to contribute to the war effort and he joined the metallurgical research staff at Watertown Arsenal. It wasn't long after he arrived in Watertown that he became fed up with the incompetents in the Civil Service and vowed that as soon as the war was over he would leave. He kept his word.

Richard was aware that the two forms of pure iron were stable over different temperature intervals. The ferromagnetic or alpha form was stable from absolute zero to 910 degrees centigrade at which temperature it converted to a new crystal structure, gamma iron, and remained in this form until 1400 degrees when it reverted back to the alpha form until it melted at 1539 degrees. The atomic magnets in the alpha form become disoriented with increasing temperature such that above 770 degrees it was no longer ferromagnetic. This was called the Curie temperature, named after Marie Curie's husband Pierre.

If one could somehow stabilize gamma iron at all temperatures it might reveal some of the reasons for the two different forms. It was at this point that Richard and Ken (Tauer) realized that thermodynamics provided some of the clues since the free energies of the two forms had to be equal at the two conversion temperatures 910 degrees and 1400 degrees. The free energy was a thermodynamic property that depended on the strength of the electron bonds and on any magnetic or other electronic effects. By carefully reconstructing the properties of gamma iron so that it matched alpha iron at the two temperatures (910 and 1400) Ken and Richard concluded that something unusual was happening in gamma iron at temperatures below which it could be stabilized. It was like suddenly seeing what was behind an invisible wall by gazing into a crystal ball. This technique offered promise to bring thermodynamics into sharper focus for those interested in solid-state physics. Over the next few years the secret of gamma iron would be unlocked and thermodynamic analysis would proliferate and eventually attract enough aficionados to warrant a journal of its own (CALPHAD, Larry Kaufman, founder and editor).

Physics reared itself at the oddest times. Richard realized this when he was punting on the Cam and the long pole became mired on the river bottom, upsetting his balance. Gravity took hold; Richard fell in, touched the shallow bottom, and bobbed up, his quenched cigar still clenched between his teeth. Momentum and thermodynamics rapidly influenced his course of action as he hastened to get home and change. From November on he shivered his way through months of bitter cold in an unheated laboratory with a space heater provided him for the colder months. He placed this directly under his chair but it did little to erase the chills his upper body suffered.

With teeth chattering Richard read a letter to the editor in the Guardian from a schoolteacher who facetiously advised how to encourage the common cold:

"A campaign against the common cold should be started in our schools where heating and ventilation are bad.

At the beginning of the school day the windows of the classroom must be kept closed until the recommended temperature of 56 degrees Fahrenheit is reached. By that time the teacher is too busy to struggle with the window pole to open the window. The cold germs then have a wonderful time in our classroom and few escape!"

Any country whose children could survive a 56-degree or less ambient temperature deserved Richard's admiration.

When spring finally arrived Richard wandered along the crocus covered 'backs', i.e. the open fields behind the river Cam; he still drank tea, and more tea, to keep warm; he boarded the steam-driven trains to London to visit Imperial College, the museums, and the American Embassy in Grosvener Square, and even attempted an impossible train journey to Oxford. The Embassy was housed in an elegant old building, later replaced with a new structure. It maintained a liquor store where duty-free and tax-free spirits (Scotch excepted) could be purchased by US Defense Department personnel for a pittance. When Richard entertained English friends he'd offer them 100 proof bourbon and sit back to see their response when the alcohol singed their tongue.

Student pranks at Cambridge were no different than elsewhere, perhaps more clever. Richard enjoyed the caper some of them perpetrated when they placed a small van on the tall roof of the Senate House, all in one night! The municipal authorities were buffaloed and had to resort to a cutting torch to dismantle it and then lower the pieces by crane, holding up traffic for hours in those narrow streets. How had they accomplished this? The students had purchased a junked van, cut it down into small sections and carried these pieces to the roof. There, the sections were bolted together in such a way that the bolts were hidden from view.

Without a car British Rail became a way of life for Richard. A glance at the rail system map revealed a multi-spoke cartwheel with London at the hub. To and from London was easy but travel between spokes, such as the 3 hours to cover the 50 miles from Cambridge to Oxford, would only be undertaken by the foolhardy adventurer. The most charming aspect of this journey between the two college towns was the gas lit station where the stationmaster extinguished the flame at 9:30 PM as the last train passed though. A few years later the Government engaged a physicist, Richard Beeching, to make the railroads more efficient. He closed down about ¼ of the lines. After that, the word 'Beechinged' entered the vocabulary, signifying a railroad line unsympathetically made redundant. Richard wrote a TV play based on such a quaint rail station but could not get it produced. He converted it into a book but the English literary agent did not see the humor. Richard concluded that the British took their railroading too seriously; George and Robert Stephenson were their heroes.

The day following his arrival in London Richard hailed a taxi and asked in his Bronx accent to be taken to the station — the cabby was amused. There were about 13 train stations and trains left from more than one station for identical destinations.

"Where you goin', guv'ner?"

"Cambridge."

"Right, guv'."

Guv'ner or guv' was a uniquely English salutation that established the pecking order between individuals. Dropped at Liverpool Street Station with his portfolio, Richard asked for the ticket window although he should have requested the booking office. Richard queued up, a passionate folkway amongst the

English to prove they were the most civilized of peoples. He requested a second-class ticket to Cambridge only to be advised there was but first and third. Richard noted a sign advertising a cheap day return to Cambridge for 8/3 i.e. eight shillings three pence.

"One way to Cambridge." (He should have requested a single).

"That will be 12 shillings."

"I can make a round trip for 8/3 and must pay 12 shillings for one way, sorry, single?"

The ticket seller hesitated, "No luggage allowed on the cheap day return."

"Is this portfolio luggage?" asked Richard.

Hesitation. "Very well 8/3."

The vapor filled station reminded Richard of the famous painting by Monet of a train station in Paris — the artist was more enamored of the steam than the station. As the train pulled into the platform Richard was struck by the appearance of elbows protruding from each door. One had to lower the window from the inside, stick one's arm out and grasp the outer door handle — it could only be opened from the outside. Richard boarded a typical corridor train, so often seen in English films, and entered a compartment already containing several passengers. There was nary a 'hello' nor a smile — Richard was a stranger. Heaven forbid they should discover he was an American, considering there must have been half a dozen items of his apparel that would have identified him. This reaction paralleled pub behavior where a stranger was viewed with silent suspicion upon entering. However, on leaving, everyone would bid him goodnight.

Britain was a country of rail enthusiasts, partly explained by the world's first introduction of railroads in the north of England in 1825. George Stephenson of Newcastle pioneered their development, a man with no formal training but with a natural bent for the intricacies of mechanical devices. In 1956 most distances traveled in England were too short to warrant air travel so that the railroads remained a vital part of the transportation economy. Over a million commuters were carried in and out of London every day. The awesome bulk and rasping breath of the locomotives, their long steam entrails and frequent showers of soot, the unsteady accelerations and deceleration, and the beige walls of the tea rooms, unpainted for years, all seemed a 25 year step backward in time. Still, Richard found it unbelievably romantic and it probably accounted for his nascent yearning to recapture the past.

Pubs in Cambridge were packed at lunchtime when university staff retired thereto to imbibe their ale (Richard fancied Flowers' Bitter), have a pub lunch like a Ploughman's (French bread, butter, cheddar cheese, an onion and chutney), and chatter away (bring your own banter). Pub ambience became very much ingrained in Richard's psyche to be resurrected some twenty years later when he became a 'colonial' tavern keeper in Massachusetts.

In the fall of 1956 it was decided to honor the memory of J.J. Thomson, Cavendish Professor preceding Ernest Rutherford. On Saturday, December 15, 1956 Richard attended a Commemoration Dinner at the Cavendish Lab celebrating the hundredth anniversary of the birth of Sir J.J. Thomson, Nobel Laureate in 1906 for the discovery of the electron. Black tie, of course. Off to the tailors Moss Bros. to rent a tuxedo. What excitement!

By the time of the Commemoration Dinner Richard had had several months to get the hang of handling the knife and the fork (fork always in the left hand even when eating peas) but he realized it would take years of training to learn his wines. The seven course dinner — consommé, filet of sole, roast duckling, fruit, dessert, breads, and coffee — was enhanced with an Amontillado, a 1952 Beaujolais, a 1949 Zeltinger, a Croft 1945 dessert wine, and a Sercial. Discussion at the meal about the wines left Richard back in the Bronx thinking about coca cola. (Actually the Navy had turned him into a scotch drinker, wine was considered too slow for a tar on short leave.) There were three toasts that evening; the mandatory one to the young Queen, one to the memory of Sir J.J. Thomson by the Master of Trinity College, and one to the Cavendish Laboratory by Sir John Cockcroft (inventor of the linear accelerator) with a response by Cavendish Professor Nevill Mott.

At the beginning of the ceremonial toasts the stewards filled the wine glasses, beginning with the head table. As soon as Mott's glass was topped up he arose to toast the Queen but he failed to realize that the other tables were still dry. An embarrassed Cavendish Professor sat, chuckled, and waited for the wine to be served to all. Mott fast became a legend as the absent-minded professor. He had assumed the post at Cavendish Lab after his tenure at the University of Bristol. One day a month he'd scoot off to Harwell for a day of consulting. Shortly after assuming his new position at Cambridge he eschewed the train and drove to Harwell with Mrs. Mott, leaving her in Reading for a day of shopping. At 5 PM he was asked by the secretary at Harwell if he needed his usual ride to the station and without batting an eye replied in the affirmative. Before he awakened to the reality of his new life at Cambridge he was on a train heading west to Bristol, his wife was south at Reading wondering why he was late, his car was east in the parking lot at Harwell, and his job was north at Cavendish Lab! Fortunately absent-mindedness did not deter the Nobel Committee from awarding him the Prize in 1977.

A few weeks after settling in at Cavendish Richard met Professor Douglas Hartree, the physicist who had distinguished himself with his calculations of electron distributions on atoms. He had developed an approach to the problem that set the standard for years to come. His father, a mathematician looking for something to do after he retired, began to help his son in making these calculations. Their approach was simple, straightforward, but toilsome.

After Schroedinger discovered the quantum mechanical equation bearing his name (1926) it was employed by theoreticians to fashion a mathematical approach for calculating electron distributions on atoms, i.e. how far from the nucleus did the electrons wander on an atom? For the simplest of all atoms, hydrogen with but one electron, the equation could be solved exactly and the answers for the energies agreed precisely with observations in the laboratory. The manner in which the electron moved around the nucleus was also readily reduced to simple equations although no one had been able to confirm this in the laboratory since the experiment was too difficult.

Hartree entered the picture when he set out to determine the approximate answers for the energies and electron distributions for atoms with more than one electron. Complications arose in these cases because the negatively charged electrons repelled each other and simple solutions to the Schroedinger Equation were no longer possible. Hartree developed a procedure whereby he'd guess the answer to the electron distribution and then calculate the energy. He would then alter his guess a wee bit and try to reduce the energy. This process would continue until he achieved the lowest energy he could. Since in reality the electrons arranged themselves to keep the energy as low as possible, Hartree's approximate answer would provide the physicist with an estimated solution. This may sound like a Rube Goldberg idea but with nothing better, you made do with it.

The calculated Hartree energies were within 1/4% of the experimental energies but since no one had measured the electron distributions for atoms one could not compare Hartree's calculation with experiment. Richard had measured the electron distribution of iron but in the form of the metal. One did not expect the electrons on the isolated atom to be the same as in the solid. So there stood Richard attempting to reconcile apples and oranges, Hartree's isolated atom of iron with Richard's experimental measurements on iron atoms in the metal and, according to Richard's results, they differed considerably!

While attempting the next step Richard turned to acting — when the real world was overburdensome one tried some form of escape. Cambridge boasted at least one theater for stage plays and The Compass Club Director decided on the American play *OUR TOWN* by Thornton Wilder. He needed an American for the principal role of the stage manager, a narrator to the on-stage events who would provide additional commentary at the intervals. The play opened on March 28, 1957 and the Cambridge Daily

News reviewed it the following day in the same issue that reported the theft of 20 iron manhole covers valued at £5 each. Ironic, thought Richard. To slightly paraphrase the review:

> The Compass Club was bold enough to perform a play recently discussed amongst television addicts.
>
> Thornton Wilder's drama without scenery managed to hold the attention of the audience for just its two hours' duration; an extra few minutes would have been too long.
>
> It is not an easy play to perform convincingly and the Compass Club is to be commended on its ambitious choice.
>
> Of course the absence of scenery and resultant miming actions are irritating for a while, but the story, explained by the stage manager, is sufficiently absorbing to dispel such aggravation.
>
> The stage represents a small town in New Hampshire and events are introduced and commented upon by the industrious stage manager Dick Weiss, the most striking personality of the evening.
>
> Apart from the fluid natural performance of the stage manager, etc., etc.

One of Richard's colleagues, in the lobby for the interval, overheard the reviewer confiding to his companion. What really impressed him was the marvelous American accent the stage manager achieved! The Bronx had made its impact in far off England.

Before Richard left for England in 1956 Slater had informed him that he was planning a conference at MIT for the following summer and he invited Richard to present his work on the 3d electrons in iron (and the neighboring elements in the periodic table). The 3d identification referred to those electrons capable of producing magnetism. While Richard was unhappy that no one had independently confirmed his experimental results he welcomed the opportunity to have a distinguished audience assess his findings. The conference was to be entitled *CURRENT PROBLEMS IN CRYSTAL PHYSICS.* Richard accepted the invitation and wrote the director at Watertown from England requesting travel orders to return to Boston for a week in order to deliver his paper at the MIT conference. The surprising reply was in the negative — Richard's request was not considered part of his mission as a Rockefeller recipient slated to study abroad! Bill Cochran, an x-ray experimentalist at Cavendish was also invited to the conference and he agreed to deliver Richard's paper. An impressive list of presenters included W.H. Taylor, D. Hartree, R.J. Elliott, R.D. Lowde, George Bacon, Helen Megaw, R. Pepinsky, H.A. Levy, B. Lax, Dame Kathleen Lonsdale, H. Curien, and B. Brockhouse. The total attendance was about 150. Richard heard that Cochran's presentation of his paper led to a lengthy discussion but no one put forward convincing reasons to shoot his work down!

Not to be outdone, a private conference entitled *ELECTRON THEORY OF TRANSITION METALS* was planned by Nevill Mott for June at the Cavendish Lab. It was attended by the theoreticians H. Jones, Harvey Brooks, R.J. Elliott, Longuet-Higgins, E. Wohlfarth, J. Friedel, B. Coles, K.W. Stevens, L. Orgel, W. Lomer, Walter Marshall, J.S. Griffiths, P. de Gennes, S. Altmann, Mlle. Galula, and R. Weiss. Discussion about Richard's results was picked up by the group and after a certain amount of post-operative analysis the theoreticians were able to convince themselves that Richard's results did, indeed, fall into nature's scheme of things. The idea was slowly being drawn into the transition metal folklore by the rationalization of the theoreticians. Still, with no experimentalist to confirm or challenge Richard's results they could do little else than shoot Richard or try to explain the answer. Some wished they had opted for the former!

Actually, there was a bizarre turn-of-the-century precedent that Richard knew had occurred in France after the discovery of x-rays by Roentgen. A French physicist announced the follow-up discovery of N-rays but, unlike Richard's results, over a hundred other Frenchmen reported experimental confirmation. However, no non-French scientist could find these N-rays. The famous American physicist R.W. Wood visited the laboratory where the N-rays were discovered and performed some sleight-of-hand in the dark to remove a critical component of the experiment. The observation of N-rays continued unabated but when Wood reported his 'dastardly act' in an English journal the bubble burst. Mass psychology was as effective amongst French experimental physicists as amongst Anglo Saxon theoreticians faced with Richard's results.

Some practical but inexplicable thermodynamics not covered in Zemansky's book reared its head. Richard spent the winter in a house in Cambridge, which he rented from an academic in philosophy who was off to the Near East to enjoy its warmth and to brush up on his scholarly pursuits. On his return he was shocked to discover that many of his philosophy books had developed a coat of mildew. When winter was afoot Richard did what he considered necessary to survive — he turned up the heat and closed the windows.

"Don't you know you must keep the windows open for proper circulation?" posed the philosopher.

"Don't you know drafts can be fatal to one's constitution?" retorted Richard.

"You've ruined my set of philosophy books."

"Why didn't you inform me what to do before you left?"

"I should not have to tell you the obvious."

And so it came to pass that a bill for £200 was presented to Richard by the philosopher's lawyer or barrister or whatever he was called. Not to give Americans a bad name Richard paid it and even threw in an apology.

W.H. Taylor went off to America for a conference and a bit of tourism. His Canterbridgian accent (i.e. his precise Cambridge pronunciation) followed him to Boston where he picked up Richard's Studebaker and toured the country. In Texas he was stopped by a patrol car and, when the officer approached him, Taylor asked him the nature of his offense. The officer screwed up his face in bewilderment and walked to the front of the car to inspect the license plate. He returned with a smile and addressed Taylor,

"Oh, you'all is from Massachusetts."

Walter Marshall told Richard about an encounter he had with a couple he met in Texas who asked where he was from.

"England," he replied.

"Did you drive all the way?"

The Suez crisis reared up midway through Richard's Cambridge sojourn. This so aroused the ire of the English that on one occasion Richard was thrown out of a greengrocer he patronized. Even though it was 183 years Richard felt that the Boston Tea Party should have left some impression — the Americans were not to be trifled with. It may take a year abroad to make an objective evaluation of oneself as an American. Still, it was a bit harsh denying Richard a head of cabbage over the Suez Canal. He was hardly in a position to disagree with the dictates of President Eisenhower.

To complete his Rockefeller mission Richard convinced the authorities at Watertown that he required a 90-day continental tour. Orders were written for him to proceed on or about 23 July from Cambridge to visit Frankfurt, Copenhagen, Stockholm, Brussels, Paris, Zurich, and any other place he desired enroute. All paid for! So off he went on his tour, ready to talk to anyone willing to be enlightened about the 3d electrons in iron. Richard had been converted into an officer and a gentleman in 90 days at Annapolis, could he convert a few European physicists into believers about iron in but a few days?

Before he left for the continent he had to bite the bullet. So far all his work on iron was hearsay. But now Slater wanted to publish the conference papers in the Reviews of Modern Physics, an important technical journal. There would be no turning back once Richard's results appeared in print. It was 3 years since he first began making noises about the low number of 3d electrons in iron; it was time to stand up and be counted. Lomer and Marshall had already submitted a theoretical paper to the Philosophical Magazine explaining Richard's x-ray results with a footnote stating that the Weiss results were to be published. Richard had placed himself in a corner. It was evident that no one was going to jump in and repeat his measurements — he was on his own. 'Publish or perish', an oft-quoted phrase in Academia, might soon be replaced by 'publish and perish'.

During this time Douglas Hartree had been most sympathetic to Richard's plight in not having any experimental confirmation of his results. When asked whether the results were reasonable Hartree could not offer a response. What happened in a metal where each iron atom was surrounded by eight neighboring atoms with which it shared electrons was impossible for him to say. But whatever ensued Richard felt certain that Hartee would try to be as helpful as he could. Alas while waiting to board a bus he dropped dead of a heart attack!

Richard purchased a used Hillman auto, with the left hand drive, from a G.I. and headed off to the continent to meet with a number of professors he had contacted by post. He arrived at Dover to find a large group of American college students traveling en masse ready to board the ferry to Finland. As each person boarded the boat the immigration authorities recorded their identities and passport numbers. Their casual dress and knapsacks left a permanent impression with Richard who later read about the meeting in Finland. This accidental encounter would haunt Richard a few years later since the international 'Peace Conference' was organized by a left wing group. Richard was mistakenly identified as one of them!

The technical purpose of his trip was a flop. Richard's unusual results for the distribution of electrons in iron had scarcely reached the continent. The ravages of war were still being cleaned up; the economic recovery in all countries could barely support research programs. Priority had to be given to re-establishing the universities. Little interest was shown in listening to Richard — scientists were more interested in talking about their own work.

At the government research center south of Paris Richard was shown a historically intriguing piece of equipment. Pierre Weiss' magnet employed in the early research on the strength of magnetism in iron, nickel, cobalt, etc. Pierre Weiss, an Alsatian working in Strasburg before the war, had established the unit of magnetism. Just before the Germans moved into Strasburg the magnet was moved to Paris and hidden although it was not clear how the magnet could have aided the German war effort. Ever since the Curies had worked on magnetism a tradition for such research developed in France culminating in the Nobel Prize work of Louis Néel on antiferromagnetism.

It was a marvelous trip with stops in Brussels where he was introduced to le mannequin Pis, a small cherub of a statue employing nature's way to water a fountain. The statue possessed a large wardrobe of military uniforms fabricated by the various Allied regiments that passed through Brussels. In Mastricht, Holland, he was guided through the limestone caves that were used to store important Dutch paintings, like Rembrandt's *Night Watch*, to keep them from being confiscated by Göring. The Germans were too apprehensive to wander through the myriad of limestone tunnels for fear of getting lost. The guide spoke highly of the cans of American Spam that were eagerly consumed during and after the war. G.I.'s and Richard hated the stuff. In Copenhagen he visited the Carlsberg brewery well known for supporting theoretical physics. The Niels Bohr Institute was built behind the brewery and to this day is supported by Carlsberg profits. 'Drink Carlsberg and support theoretical physics!'

When in Copenhagen he did not miss the Tivoli Gardens a unique combination of posh cafes and sideshow entertainers. It expressed the very essence of Danish culture. It was the only time Richard had an opportunity to see trained fleas. They were hitched up to small carts and on command would pull them along.

Richard took the ferry from Helsingor, the site of the Hamlet Castle in Denmark, to Sweden where he turned north toward Stockholm. He had barely traveled 20 miles when he came upon a tall gaunt figure of about 50 walking the road and carrying two pieces of luggage. He was thumbing a ride and Richard stopped but, after noticing the US Armed Forces license plate, he was reluctant to accept Richard's offer of a lift.

"Ich bin Deutsch," he said.

Richard smiled and assured him it did not matter. The cadaverous figure was elated and informed Richard he had but 5 Swedish Crowns to get to Stockholm. They stopped at a hotel and Richard treated him to a meal. What luck that Santa Claus came along! Furthermore, he told Richard, he was fleeing East Germany (four years before the wall went up) and heading to Stockholm to visit his aunt. In Richard's poor German he managed to learn how terrible conditions were in East Germany. Karl had fought in the war but German veterans were not granted pensions, their wounded were given minimal care, and their elderly had to rely on their children's help.

When they reached the aunt's address in Stockholm Richard bid him good luck.

"Is your aunt expecting you?"

"Nein."

"When was the last time you saw her?"

"When I was five."

"Fifty years. Will she remember you?"

He shrugged and moved off.

In Upsalla, a notable university town north of Stockholm, Richard met Professor Kai Siegbahn a member of the Nobel committee. He assured Richard that all candidates for the Prize were given a thorough background check. Not only were their research papers carefully studied but aspects of a candidate's personal life were examined. It was de rigueur that the candidate must not campaign for the Prize and his personal life had to be devoid of immoral behavior.

In Norway he had to navigate the winding roads. He asked why no signs were posted to warn motorists of bends in the road,

"They only post turns of more than 120 degrees!" he was informed, "Otherwise there would be more signs than trees."

Back in Cambridge Richard packed for the trip home. He wondered what sort of reaction there would be after his results on iron appeared in print.

WATERTOWN 2: 1958–1961

The papers presented at Slater's conference were published in the Reviews of Modern Physics in January 1958. In a short report of but 4 pages Weiss and DeMarco outlined the experimental approach to the determination of the x-ray scattering in chromium, iron, cobalt, nickel, and copper metals. Critical to the measurements was the correction for extinction (multiple scattering) by making the measurements at several wavelengths. The first two of these metals crystallized in a form such that each atom had eight near neighbors while the latter three crystallized in a structure such that each atom had 12 near neighbors. The x-ray results were compared to the isolated atom calculations of the Hartrees and revealed that cobalt, nickel, and copper did not differ but iron and chromium varied considerably between isolated atom and metal. One argument put forward in the Weiss-DeMarco paper was that all five metals were measured in the same manner. If the results were trustworthy for some they should be reliable for all. The number of 3d electrons for each metal were:

Number of 3d Electrons

Element	Isolated Atom	Measurement
Copper	10	9.8 + 0.3
Nickel	9	8.7 + 0.3
Cobalt	8	8.4 + 0.3
Iron	7	2.3 + 0.3
Chromium	5	0.2 + 0.4

It appeared that iron and chromium had undergone a significant reduction in the number of 3d electrons between isolated atom and metal. How could this be? Where had these electrons gone? Was it just the crystalline arrangement? Some theoreticians thought this to be the explanation and published their ideas to underscore this. They actually went out on a limb with their interpretations by resorting to hocus-pocus rationalization.

One such theoretician Walter Marshall (later Lord Marshall for his work on overseeing the power reactor program in England) jokingly advised Richard that if his measurements were faulty he'd better find a job in Tahiti.

Amongst the theoreticians Richard encountered, many accepted the experimental results, only two were firm in their disbelief, while a large group chose not to get involved and wisely remained noncommittal. And there the matter stood for a few years!

Dave Chipman had now become a player in this scientific mystery when he joined the Watertown group around 1956. An outstanding student of Bert Warren's and son of a metallurgy Professor at MIT, he teamed up with DeMarco while Weiss was at Cavendish Lab and the pair made some efforts to repeat the measurements on the same sample of iron previously used. Little more could be added to the previous measurements.

All work and no play made a physicist a longhair! Dave was a dedicated sailor and a number of his scientific colleagues pooled their resources to hire a sailboat for the America's Cup races off Newport. The races had been postponed due to the war but in September 1958 the British challenged with their new boat Sceptre. The defender, Columbia, from the New York Yacht Club easily beat the challenger. Richard was impressed with the sight of thousands of spectator boats gathered some twenty miles from Newport RI. It reminded him of the Eighth Fleet amassed before the battle of Okinawa when the entire horizon was filled with seagoing vessels.

At the end of the race Dave steered the boat to get a close peek at the Sceptre. With several of the scientists standing in the bow obstructing his view the boat narrowly averted ramming the Sceptre. Fortunately the British crew came forward to ward off the 'crazy' Yanks. With millions of dollars having gone into its construction such a collision would have blared from the headlines and made it impossible for this crew of physicists to live down.

Richard had returned from Cambridge, England, at the end of 1957 with some of the Bronx excised from him. He was anxious to get back to his x-ray measurements but he was short of ideas as to the next step. He needed some diversion and fancied the idea of seriously pursuing theater as a hobby although not as an amateur. As nearly as he could ascertain amateur actors not only lacked competence but they were unable to assess how poorly they performed. The story is told of the lass with an actor boyfriend whose father would not grant permission for her to wed a thespian. She implored her father to see him act before he made up his mind. The father returned from viewing a performance and announced to his daughter,

"You can marry him — he's no actor."

Richard made contact with the Actor's Workshop in Boston and met its director Alan Levitt, a personality who exerted as much influence as Slater in stimulating in him an interest in a subject. Alan had been a product of Emerson College and had studied in France with the renowned Jean Louis Barrault. He would later go off to Hollywood as a successful writer for the popular 'All in the Family', receiving an Emmy nomination for one of his shows. Alan and Richard became good friends.

All the students at the Workshop including Olympia Dukakis were studying acting but Richard asked to be taken on as a directing student. Alan indicated he did not know how to teach directing but agreed to try. Working with two or three acting students at a time, Richard would rehearse short scenes with them before presenting it as a class exercise that Alan would critique. It soon became apparent to Richard that acting and physics represented diametrically opposite approaches to their subjects. In physics it was essential that unemotional objectivity be maintained in evaluating experimental results. If one allowed personal feelings to impose themselves one could easily be swayed by marginal data or observations. The cold, hard facts had to be distilled without personal bias.

On the other hand, plays are almost wholly devoted to revealing the inner emotions of the actors. It's self-defeating trying to be totally analytical about a role. This dichotomy in approaching the two subjects may account for the fact that very few physicists take up acting or playwriting as a hobby. Contrast this with the mathematical regularity in Bach's music that lures the physicist to classical music. They are found in great numbers as both players and listeners.

In dealing with a play the director must first ascertain what the playwright had in mind based solely on the dialogue. In any scene actors want something from each other and a good playwright injects these hidden motives into his dialogue. If the director and actor understand what these are then it is the role of the director to help the actors find the internal and external route toward achieving this end.

In science a critical ingredient in establishing veracity was reproducibility, i.e. the measurement must produce the same answer from day to day and from one lab to another. This was a necessary but not sufficient condition for it was possible to repeat the same mistake every day but less likely when an independent measurement was made in another laboratory. Hence confirmation was mandatory for scientific acceptance. The theater, on the other hand, permitted considerable leeway in interpretation and one rarely saw two productions of a play performed in an identical manner.

During Richard's tenure at Brookhaven, when he shared an office with Si Pasternak, assistant editor of the Physical Review, a paper was submitted by two competent mathematicians from the University of Pennsylvania. They had repeated the Duke University experiment in which student 'guessed' the turn of a card. The group had come up with a statistically significant positive effect. The referees could find nothing wrong with the experiment — should the leading physics journal publish the paper? The editors decided against acceptance since this would give credence to psycho kinesis, the ability to exercise 'thought' in the outcome of an event. That would undermine the repro-ducibility requirement for approval in science. After all, every experimentalist has some preconceived idea how an experiment will turn out. If his brain images are affecting the results we become hopelessly mired in psychology. Science has adopted the rule about reproducibility; firm in the belief that nature is not fickle and will not let them down.

One year after Richard entered the Actors Workshop Alan Levitt decided to stage Chekhov's "Cherry Orchard' and enlisted Richard as stage manager, an exacting task in looking after props, etc. for a large cast. The 'Cherry Orchard' was not an easy play to perform partly due to the difficulty in obtaining a good translation. The nuances in Russian are sometimes impossible to express in English. Many of Richard's scientific colleagues showed up for the performances but there was very little agreement amongst them about the interpretation. Chekhov was also troublesome because the actors stood around complaining about life and did very little about it. Scientists were generally bored with Chekhov since nothing seemed to be happening.

Naturally, as Richard became more absorbed in plays his role as a physicist and as a student of directing prompted him to consider regions of overlap between the two. In both science and theater one encountered an audience of listeners and a presenter. The theater had been much more successful in keeping audiences awake even though they both had a story line to propound. Scientists were not trained to give dramatically interesting performances yet there was a parallel, i.e. both actor and scientist must start with a script that had something to tell and both must deliver the script in an absorbing manner. Furthermore, thought Richard, if actors can be trained to handle themselves on stage why not scientists? Of course a physicist was not simply providing entertainment, he was trying to present ideas about nature that would help the audience in his or her own research. If a playwright was successful the audience would leave the theater with a feeling of elation and a few thoughts that might affect their own being. This sort of analysis filled Richard's thoughts over the years.

The Stanford Research Institute was the recipient of research grants to fund work on the proton accelerator they had inherited from Watertown in the debacle over the 'pair' of van de Graaff accelerators. Watertown had the responsibility for monitoring a quarter million dollar research program on impurities in gun barrel steel. Since nuclear physics played a dominant role in the work Richard was asked to assess the experiment. He flew out to San Francisco and was chauffeured to Stanford. After visiting the laboratory and speaking to the principals he returned to Watertown fairly unimpressed. He expressed his disappointment and went on to blurt out,

"For $250,000 you could buy a small reactor."

Those words set in motion a process that would excite the Army Generals in Ordnance. They asked Richard to prepare a proposal on Watertown's need for a reactor. Richard complied in the belief that the universities and graduate students in the Cambridge area would supplement Watertown's own research program. Watertown alone could not make efficient use of a tool that accommodated 50 to 100 researchers.

It was necessary for the Army's request to snake its way through the chain of command. At this time the Naval Research Lab in Washington entertained an identical idea and the two branches of the service clashed over the matter. Whoever decided priorities in the Defense Department gave it to the Navy first with the promise that the Army would get a reactor in the next budget. Without waiting for the next congressional authorization Richard was told to become involved in developing this project. He met on a regular basis with a reactor design company in Detroit to ensure that the experimental area around the Watertown reactor would be uncluttered with machinery, etc. it was to be a dedicated research tool for neutron diffraction and there'd have to be plenty of space for the experimentalists.

Richard found Detroit to be a forgettable city although he was pleased to discover a Chinese restaurant that served pressed duck, fairly rare to stumble upon. Returning from one of these two-day trips Richard was ordered to get down to Washington with all his notes about the reactor. He was not told why although he realized that something important was brewing. The next day he and the Commanding Officer of Watertown caught the morning flight to Washington National Airport and by noon were closeted in the Pentagon with the austere General Advisory Committee. Oppenheimer and Fermi were principal members and Richard began to go over in his mind the presentation he would make to convince them that the Army needed a reactor to do neutron diffraction. He admitted to himself that he could not plead any stronger case than the time Shockley at Bell Labs asked what he could do with neutrons. In spite of his theatrical training Richard knew he had a poor script and trepidation set in. Richard had visions of making an unconvincing presentation and forever being accused of losing the program! Oppenheimer could easily shoot down the project with one of his caustic barbs.

Richard and the Colonel stood at the rear of the room waiting to be called. When the Watertown project was announced someone from Army Ordnance in Washington said a few words. Richard's knees were vibrating as he waited in the wings for his cue. He was scarcely cognizant of the events taking place. Five minutes later he and the Colonel were led out of the room, the Watertown reactor had been approved without a word from Richard!

Prior to leaving Cavendish Lab one of the crystallographers had approached Richard on behalf of Rosalind Franklin who was scheduled to make a trip to Boston. Money was very tight in the UK, could Richard find her a place to stay? Richard agreed she could stay at his place. He was then introduced to Rosalind and they parted company.

The lady crystallographer who had provided the crucial x-ray data in the DNA discovery turned up on schedule in Boston and Richard looked after her for several days. Pleasant-looking, with very little effort at makeup, dark hair uncoiffeured, she had made her way in a man's world by a degree of reticence and to a great extent by working alone. It was customary when scientists met that they would discuss their work but when Richard casually posed this question Rosalind clammed up and never once asked Richard about his own work. Odd, he thought. A few years later James Watson wrote his book *THE DOUBLE HELIX*, an account of the discovery of DNA. In it he was unduly critical of Rosalind and aroused the ire of many scientists. She may well have put James in his place at one time. Since she died before the 1962 Nobel Awards were bestowed she became ineligible, but in retrospect she was as responsible as Watson, Crick, and Wilkins for the discovery.

When the BBC did a full-scale dramatization of the story with Jeff Goldblum portraying James Watson, they presented the character in a most uncomplimentary way. They felt he deserved it — a remark the producer privately confided to Richard.

Just before Watertown awarded the contract for the construction of the reactor, MIT announced they had been granted a gift to build a reactor for research and teaching. With the coming atomic age and the expected proliferation of nuclear power technology MIT felt it needed a reactor to remain at the forefront of the needs of the academic community. At this point Richard expressed some concern to the Commanding Officer — two research reactors only a few miles apart were hardly justified. The CO agreed and the two discussed the matter with the MIT Dean of Engineering.

The Colonel offered to enter into a cooperative arrangement with MIT to construct a single reactor for research purposes. This was turned down. MIT had had enough experience with the government. Their sponsors had provided adequate funding — money was not the issue. The Colonel pleaded but it was clear MIT insisted on going it alone. However, the Colonel was assured that Watertown would have access to the MIT reactor and with some encouragement from Richard it was agreed that this appeared to be a practical solution. It was unlikely that more than 3 or 4 Watertown scientists would be doing neutron diffraction.

And so it came to pass in the kooky world of government bureaucracy that the idea was shot down by the Washington brass. Watertown was going to have a reactor whether it liked it or not! No doubt the brass didn't wish to appear ridiculous in the eyes of

Congress. Having requested the reactor how could they now request abandonment of the project? Didn't the Army know what they wanted? One Colonel assured an audience that the reactor was needed to look for the 3d electrons, the very words had come back to haunt Richard!

On May 17, 1960 the Horace Hardy Lester reactor was commissioned only to be abandoned twenty years later at a dismantling cost many times the original price of construction. What else is new?

Early in 1959 Professor Wolf in the Metallurgy Department at MIT asked Richard to teach his course while he was away on sabbatical. Richard agreed and received an official notification on February 10, 1959 as Lecturer in Metallurgy at a salary of <u>no</u> dollars to be paid in monthly installments. The last time Richard did any teaching was at the high school in Oakland as part of his University of California graduate course in Education.

That had turned out to be a disaster since Richard found no interest amongst the students to learn anything. But this was MIT — it would be different! Alas! The students were only concerned with obtaining a good grade at minimum effort. Hence, Richard's informal approach with very few notes was viewed with suspicion. The experience further slammed shut any interest in teaching and Richard gladly returned to his experiments at Watertown. Advice was only worth what MIT students would pay for it.

Physicists, like actors, become prima donnas when they were elevated above their fellow man, the greater the height the more haughty and arrogant. As they say, 'nice guys finish last'. Some of these characteristics were displayed between physicists but less so between scientist and the lay public. Sometimes the 'chemistry' between otherwise normal physicists stirred up a personality conflict that could not be confined. So it was between Arthur Freeman and Bob Nathans — and Richard was the accidental catalyst.

It started with a letter from W.H. Taylor:

UNIVERSITY OF CAMBRIDGE DEPARTMENT OF PHYSICS
Cavendish Laboratory
Free School Lane
Cambridge

Dr. R.J. Weiss
Ordnance Materials Research Office,
Watertown Arsenal,
Watertown, 72,
Mass. USA

Dear Dick,

I send you herewith the typescript of a short note, which will shortly appear in Philosophical Magazine. As you will see, we have been mildly alarmed by the cheerful way in which the theoreticians seize on numerical results without, in some cases, stopping to consider under what conditions particular measurements can be generalized.

Yours sincerely,
s/Will
W. H. Taylor

The paper was a joint contribution from W. Hume-Rothery (Department of Metallurgy, Oxford) and P.J. Brown, B. Forsyth, and W.H. Taylor from Cavendish. Hume-Rothery was a well-known metallurgist who was remembered for his total deafness. Whenever Richard discussed science with him, Hume-Rothery would hand him a blank pad and pencil and ask Richard to write everything down. Richard, as well as others, found himself 'psyched out' by this procedure.

For one thing it took much longer to converse this way, particularly if you did not write clearly, for then Hume-Rothery would question one's handwriting. As a second aspect of one-upmanship Hume-Rothery could sometimes refer back to a previous page of your writing and exclaim,

"But that's what you said on page 6!"

And there it was indelibly recorded in your handwriting. Richard and others tended to avoid getting into a hot scientific argument with the old boy, you never won!

And what was this quartet's line of reasoning in refuting the conclusions of the Weiss-DeMarco results? Quite simply they reasoned that the 3d electron distributions in iron and chromium were not spherical but the 3d electrons spent much more time between the near neighbors. This was not an unreasonable suggestion since the electrons that 'glue' the atoms together would be expected to occupy the territory between the near-neighbor atoms. Hume-Rothery et al. contended that his reduced the x-ray intensity and made it appear to be due to a reduction in the number of 3d electrons. They stipulated that the Weiss-DeMarco experiment was accurate but that their analysis led to an illusion.

Upon receiving this note Richard immediately conferred with Art Freeman, one of his colleagues at Watertown. Freeman and Dick Watson were two disciples of Slater who continued the work of Hartree in attempting to improve the calculations of electron distributions. Art introduced Richard to the improved theoretical techniques for doing so. Between Art and Richard a paper was prepared for the Philosophical Magazine to refute the suggestion made by Hume-Rothery et al. The effect the Oxford-Cambridge group proposed was too small in size, although they did open the door to the next step in this tale of three cities, Oxford, Cambridge and Watertown. It took six months for the Freeman-Weiss rebuttal to appear in print but when it did the Oxford-Cambridge guns were silenced. But it was only the first salvo in a war that would spread to other people and places.

While this was all taken rather seriously there were some light interludes. Walter Marshall, now visiting America spent some time at Harvard and occupied a house built by the descendents of John Hancock, a replica of the famous Hancock House on Beacon Hill that burned to the ground in 1863. Walter received a letter that suggested a novel solution as to the whereabouts of the missing 3d electrons. In effect, suggested the writer, the sample of iron studied by Weiss by directing an x-ray beam at it would be, to coin a phrase, 'shot full of holes', thus accounting for the missing 3d electrons!

About this time Richard visited Brookhaven where he was shown some neutron diffraction results on iron taken by Bob Nathans, an employee of the Navy. He had discovered some variations in the neutron intensity as the scattering angle increased. Unlike x-rays that were scattered by all electrons, neutrons were only scattered by magnetic electrons.

"Could this be due to the 3d electron distributions not being spherical?" he asked Richard.

Having just shot down this same suggestion by Hume-Rothery, Brown, Forsyth, and Taylor, Richard felt confident in refuting the possibility. Nathans gave Richard a copy of his neutron results and these were taken back to Watertown for study.

Surprise, surprise! Nathans was right. The effect was not noticeable at the smaller scattering angles in the Weiss-DeMarco x-ray measurements but Nathans had extended the neutron measurements to large scattering angles. Weiss and Freeman were able to analyze the results employing the Freeman-Watson electron distributions and they showed that the two magnetic electrons in iron took up positions in the gaps between the near neighbors. The theoretical technique for making such calculations was written up for publication and Richard called up Shull at MIT, who was a coauthor with Nathans (now on a year's leave in Japan) to suggest a joint publication. Shull acquiesced reluctantly. By rights an experimentalist had the priority to make any theoretical calculations in interpreting his measurements. The Weiss-Freeman paper, including the Nathan-Shull results was submitted to the Journal of the Physics and Chemistry of Solids. The next thing Richard knew was the sudden return of Nathans from Japan. What happened? Richard asked Nathans why he had flown back but received no reply. No doubt something had occurred between Nathans and Freeman for from that day forward the two detested each other and even appeared to enjoy their conflict.

Yet progress was definitely being made in the quest for details of electron distributions. This was the first solid evidence that the magnetic electrons in iron did not follow a spherical path. In principle, this was not a complete surprise and several facets of the physics had been established:

1. For the first time departures from spherical orbits for 3d electrons in metal was placed on a firm experimental and theoretical basis.
2. The magnetic electrons in iron avoided the near neighbors and preferred to sneak around the closest atoms, clearly a surprise.
3. The isolated atom calculations of Freeman and Watson gave good answers for the magnetic 3d electrons in iron, cobalt, and nickel.
4. Like electrons, Nathans and Freeman repelled each other. The entire physics community soon became aware of the personality conflict.

At a meeting of the American Physical Society in Cambridge, MA, Freeman and Weiss spoke to reporters and described the departure of the 3d electrons from a sphere as 'Mickey Mouse' ears. This inaccurate description perked up the interest of the fourth estate since Walt Disney had not previously been associated with physics and electrons. When this was put out to the wire services scarcely a newspaper failed to pick it up.

And now the clot thickened. A year passed following the publication of the Weiss-DeMarco experiment and Boris Batterman at Bell Labs, a former student of Bert Warren and close friend of Dave Chipman, repeated the x-ray measurements on iron and copper employing a different technique. His results showed that neither iron nor chrom were deficient in the number of 3d electrons. At last the Weiss-DeMarco measurements had created enough of a stir to have prompted someone to make further measurements. For a while Weiss and Batterman quibbled over the new results, not unexpected when two scientists disagree. But at last a fresh breeze blew through the scientific community and it broadcast a clear message to the theoreticians who had accepted the Weiss-DeMarco results as gospel, even only a minor gospel. The log jam of theoretical speculation was now broken and

Batterman, Chipman, and DeMarco (BCD) decided to combine their talents at Watertown and make their own experimental determination since by now many of the stumbling blocks had been identified.

At the end of 1959 Richard received an invitation to attend the first Australian conference on solid state physics and off he went, leaving BCD to identify and sort out the problems associated with getting the right answer. One other aspect of the uncertainty of the electron distribution in a metal like iron was the question as to whether the theoreticians could even calculate it, not just perform some hand waving arguments. Freeman and Watson were confident that they could evaluate the electron distribution for an isolated atom to about a 1% error but joining the atoms together to form a metal presented a prodigious theoretical problem that would take years and high-speed computers to reduce the error to 1%.

Before embarking on the journey 'down under' Richard made a decision that would affect his appearance for the remainder of his life. He engaged in pogonotrophy, an arcane expression for beard growing. His itinerary for the journey included visits to a number of laboratories that were recipients of research grants from the Army Research Office in Frankfurt. All told, there were about twenty grants in solid-state physics.

Richard discovered that the man in charge of this program was also named Richard Weiss with similar ethnic Hungarian background. Two Richard Weisses in the same organization! Why was the US Army engaged in this solid-state physics Marshall plan? Richard A. Weiss wrote:

> "I find it difficult to give any other answer to why the Army is supporting research in foreign countries than that it supports good research wherever it is being done. And this it does through a policy which supports the belief that such activity will ultimately redound to the credit and use of the Army as well as the country."

In making the rounds of the laboratories Richard J. encountered an interesting phenomenon. There were many laboratories that would not take US Army money. They were convinced it was tainted. For this reason the Army preferred to have only established physicists make contact with the researchers. When asked by European scientists,

"Why is the Army doing this?"

Richard would reply, "Don't look into a gift horse at the wrong end."

Richard boarded a TWA flight to San Francisco and stopped for a day to visit Berkeley and compare the campus to the earlier postwar days when he was but a graduate student (it hadn't changed noticeably). He was still 'shocked' by the disparity in architectural styles that dotted the campus. Even adjoining buildings showed a total lack of cohesiveness, demonstrating a remarkable indifference on the part of the architects. It's still the same today.

The youthful Frenchman Pierre de Gennes, to receive the Nobel Prize 30 years later, was on a sabbatical in the Berkeley Physics Department. He had become interested in the Hastings, Corliss, Weiss measurements on chromium and its odd arrangement of magnetic directions for the atoms. He showed these calculations to Richard. Charles Kittel then professor of theoretical physics at Berkeley, had confided to Richard that he thought de Gennes was the best theoretician he had ever had in his group. Prophetic. When de Gennes had spent his first postdoctoral years at Brookhaven in 1950 Richard was the first American he had ever met. He would remind Richard of this whenever they later ran into each other. Brookhaven in those days was rather isolated and Pierre and Richard socialized, but very little discussion centered about physics. Hence it came as a surprise when Kittel lauded de Gennes in such outstanding terms. Richard thought Pierre closely resembled Fernandel, the well-known French cinematic comedian, and that he exuded an equally pleasant personality. When Pierre won the Nobel Prize four decades later Richard wrote to congratulate him and asked if de Gennes still remembered him. He did.

The QANTAS flight in 1959 (originally called the Queensland and Northern Territories Air System) was one of the last propeller driven flights from the USA to Sydney and took twice as long as the jets would. Richard remembers the boredom of that journey that must have taken well over 40 hours especially after one of the engines quit en route. The steady droning of the propellers, varying periodically as the vibrations interfered acoustically, drove Richard as close to madness as he could recall. Even with refueling stops in Honolulu, Canton (Gilbert and Ellice Islands), and Fiji the flight was an experience better forgotten. Sleep would not come. The periodic flashing of the red aircraft identification light added visual annoyance to the acoustic irritation.

When the flight finally arrived in Sydney several reporters were at the airport to query this influx of scientists about the reasons for their visits. It was an unusual event for the Aussies, their first international conference. A joke made the rounds that relied on the local accent for its humor,

"Did you come to Australia to die? (today)"

The abundance of men on the smallest continent prompted,

"There are more miles than fe-miles in Australia."

Richard soon realized that Australia was like the wild west of the early American settlers; rugged individualism was a valued and necessary way of life although Victorian standards in regard to women held sway. Otherwise 'Booze, Betting, and Broads' described the principal activities of the typical male.

The July day that Richard arrived found Sydney suffering from a heat wave of over 80 degrees; this in the middle of their winter! Members of the Royal Militia, dressed in their crimson wool uniforms like the Guards at Buckingham Palace, dropped over in a dead faint. While Sydney experienced almost no snowfall, the winter temperatures ordinarily hovered near 45. The city, like San Francisco, boasted a large harbor that was entered through a narrow break in the coastline. Max Hargreaves, one of the scientists assigned to escort Richard about, pointed out the spot in the harbor chosen to build a cultural center. The Sydney Opera House, a famous landmark, would take fifteen years to complete and would require a national lottery to raise funds for its completion (it was well over budget). It arose in the shadow of another famous landmark, the Sydney Harbor Bridge. A ferry ride across the water brought one to Taronga Park and its magnificent zoo and aquarium. Richard had never seen such a wide variety of unusual sea life. It demonstrated the special place this continent held in the evolution of flora and fauna. Once attached to Asia it had drifted away in geologic times and developed its own unique species over the eons.

The fruit in Australia was varied and impressive. Richard particularly recalled the custard apple, a large succulent species best eaten with a spoon in the bathtub. It was a bit like consuming an entire uncut watermelon without getting one's ears wet.

Richard flew on to Melbourne for the conference, joining some 130 others. While he presented the neutron diffraction results of the unusual magnetic arrangement of atoms in chromium, no one asked about the 3d electrons in iron. To Richard it was just another conference except that he now sported a beard that had just passed the itching stage. Much more memorable in Melbourne was the pub popularly called 'Jimmie Watsons' which displayed a life-size painting of a nude entitled 'Chloe' inside the front entrance. You could buy a steak and cook your own on an outdoor grill. Australia was grazing country and they loved their meat.

New friends and fond memories were left behind as Richard winged his way to Singapore and over the Mideast desert to Egypt, 'A Gift of the Nile.' He stopped in Cairo and hired a driver for the day. At the pyramids he encountered a swarthy and lithe native who addressed Richard,

"Mister — the record for climbing this pyramid is four minutes, thirty-two seconds."

Richard nodded and smiled.

"And for twenty American dollars I'll break the record!"

His guide then led him to a well,

"This is the deepest well in Egypt."

The guide threw several small rocks down and it took at least five seconds to hear the splash.

"If you keep throwing rocks in, it won't remain the deepest," cautioned Richard.

Somehow the guide thought this uproariously funny and laughed loudly.

"I've been a guide for twenty years and you're the first one to warn me."

"I'm a mathematician."

The guide was at a loss to reply. Richard wondered if this man had ever met a scientist.

The tour of Europe included three labs in Italy. In Torino he looked for a pizza parlor but only found a small purveyor at the rail station. Pizza was more popular by far in America.

In Glasgow Richard found the street on which his mother had been born. Alas! All the houses had been torn down the previous week and he was unable to bring her a current photo.

Richard returned to Watertown and took up residence in the John Hancock House that had been previously occupied by Walter Marshall. The BCD work was continuing and Larry Jennings, a careful experimentalist, had been sucked into the controversy. Since the theoreticians had only performed accurate calculations on isolated atoms why not make an x-ray determination of the electron distribution on isolated atoms? This was only possible at ordinary temperatures by employing the noble gases, neon, argon, krypton, and xenon where the atoms remain far enough apart not to alter the electronic arrangement. This was a welcome and ambitious experimental enquiry that would require considerable time, patience and care.

Things were now 'jumping' within the experimental program at Watertown. Awakened to the recent recognition that the atoms were not spherical, at least the magnetic 3d electrons, Richard conceived the idea of making such measurements with x-rays. Watertown had become the experimental capital for examining electron distributions. An excitement permeated the atmosphere of their new research building and Watertown developed a reputation in solid-state physics. Funny, the physics group had still not done anything useful for the Army even though the country was now in a 'hot' cold war with Russia.

One of Richard's friends, a radiologist M.D. working for the government at the Veteran Administration's Hospital, shared an interest with Richard in the theater and, to a lesser extent, a professional interest in x-rays. Danny and Richard had often done some sailing from a dock adjoining Danny's home on Lake Cochituate. One day Danny called Richard in and, with signs of grief registering on his face, showed Richard an official letter of dismissal he had just received. The reasons for the termination harped back to his younger Depression days in New York when he had joined a left wing group. This dismissal was in the midst of the period when McCarthyism ran rampant in America. Danny, like many others during the Depression, had sought hope by association with liberal organizations. Whether the Russians were funding such organizations may be questionable but many government employees now ran scared for fear that their past naiveté would wreak havoc with their careers. Richard realized that continued association with Danny might be viewed as a security risk. Do you give up your friends? If Richard lost his clearance he would be out of a job.

As one of the crises in Richard's life this unreasonable Kafkaesque Kangaroo court of guilt by association left him feeling helpless. Danny was offered the chance to appeal but no one seemed able to have their dismissal overturned. New Yorkers were almost guilty by natal association and Richard was a Bronx boy. Danny suffered months of uncertainty, his career on the line. Fortunately he found a job in a private clinic in Ohio and successfully squirmed out of the tightening stigma of disgrace that had ruined many.

Then one day Richard was called in to see the Director at Watertown and he was turned over to an interrogator from the FBI. Following the preliminaries the questions commenced:

"Do you believe in God?" (Atheists were automatically Communists.)

"Yes."

"Are you now or have you ever been a member of the Communist Party?"

"No."

"Do you believe in the teachings of Lenin, Marx, etc., etc.?"

"No."

At this point the grilling shifted gears.

"Have you ever visited Finland? If so, when?"

Suddenly it all clicked together. That accidental encounter with the youth group boarding the ferry to France and heading for Finland socked Richard in the solar plexus. He meticulously related that event.

As a parting FBI agent cautioned,

"You must not reveal the details of this meeting to anyone not authorized to receive it."

Richard shook his hand and left. A few days later the Director at Watertown met with Richard,

"I was very surprised with what I read of the interview. Sometimes they get it wrong."

Richard thought that sometimes physicists get it wrong, too. He still didn't know how many 3d electrons there were in iron but, no matter, those electrons weren't going to be classified as security risks!

In January of 1959 Richard had a call from Professor Harvey Brooks, a consultant in anti-submarine warfare. A so-called Hunter Killer Task Force operating in the Atlantic under Admiral Thach and consisting of the aircraft carrier VALLEY FORGE, several submarines and destroyers, and an auxiliary vessel, were on duty in the Atlantic off the US east coast looking for Russian subs. They had a problem, which they thought was related to magnetism. Since Lt. Weiss was still in the Naval Reserve Harvey Brooks suggested he spend a few weeks with the Task Force to see what he thought. It wasn't until June that Richard could get away.

He flew down to Norfolk, Virginia, only to discover that the Task Force had left for sea duty. No fear, he was assured, there was a daily mail flight out to the carrier. He climbed aboard the courier plane and in less than an hour they reached the Valley Forge and made a typical carrier landing, i.e. the arresting gear on the carrier's flight deck would snare the tailhook on the plane and the aircraft would come to a rapid stop. Funny, thought Richard, three years at sea on a carrier during WWII and this was the first carrier landing he had made.

He met the tall, gray-haired Admiral Thach, a pilot during WWII who flew at Midway and in other South Pacific battles. Richard noticed through the open hatch that the Admiral slept in a real four-poster bed — the first one he had ever seen in the Navy. The Admiral indicated how they had been looking forward to Richard's visit. Like the time he first arrived at Watertown Arsenal and was told by the Commanding Officer that they were waiting for him, Richard was greeted as if he would quickly solve the Navy's problem, whatever it was.

Lieutenant Weiss was immediately assigned to one of the Admiral's aides who decided to give Richard the full treatment, two days on a submarine, a few days on the destroyer, then back to the carrier. He'd be briefed what it was all about along the way. Richard was back in the Navy — follow orders and keep your mouth shut.

The Valley Forge was the flagship for Task Force Alfa that included 7 destroyers, many airplanes for antisubmarine detection and submarine destruction, and several submarines. It was believed the Russians had almost 500 submarines although no aircraft carriers. Later on the Russians began using titanium for submarine hulls instead of steel since it was lighter and non-magnetic. One of the problems that made sonar detection of submarines difficult was the large background noise of whales, snapping shrimp, booming drumfish, squealing porpoises, and schools of fish. Finding Russian subs was more difficult than the needle in the haystack.

The day after Richard reported on board the Valley Forge an auxiliary vessel pulled alongside the carrier and with both ships steaming on parallel courses at about ten knots a breeches buoy was sent over and Richard was hauled aboard the auxiliary vessel. In a like manner he was then transferred to a destroyer. Taken to see the executive officer he was given his first inkling what it was all about. The radar on board the destroyer would occasionally be able to pick up the submarine's whereabouts from an echo above the spot where the submarine was submerged. What could it be? In a few days they would be testing it out with the submarine's cooperation. That evening Richard's thoughts were not on physics, his susceptibility to seasickness reemerged and he spent a miserable night in his bunk. Anyway, he now knew what he had been sent for.

The following day a helicopter appeared overhead, dropped a line with a tube on the end, which was placed around Richard's chest and under his armpits. Up he was hauled and placed on board. A surfaced submarine was on patrol a mile away and Richard was flown over and lowered onto its deck. He was greeted by the executive officer,

"Were you nervous?"

"It's all a matter of attitude," he replied.

For several days Richard learned about life on a crowded non-nuclear sub. When submerged there was smooth sailing. But the bunk he was given that night was so close to the outer skin of the vessel that Richard had no room to turn over!

The next day he posed the question to the CO,

"What about these radar sightings?"

Richard only received a noncommittal shrug. So far, he thought, they hardly needed a physicist.

Back to the destroyer and a seat in the radar room near the cathode ray tube. He asked the Lieutenant on duty to point out a submarine contact. When it supposedly appeared on the screen and the Lieutenant pointed it out Richard could not see a thing. And there it was — a bit like N-rays when one's imagination robbed one of his senses. Two weeks of nothingness, thought Richard. He returned to the Valley Forge and the Admiral welcomed him 'ashore'.

"What did you think?" asked Thach.

"I didn't see a thing that convinced me. I'll think about it and send you a report."

The Valley Forge returned to Norfolk. Upon his return to Watertown Richard wrote his final report. There was no way known for radar waves to travel through an electrical conductor like salt water. Instead, Richard suggested sensitive magnetometers be installed in helicopters and deviations in the earth's magnet field be measured. A large magnetic steel vessel like a submarine might show up.

To Richard's surprise the Admiral followed up this advice and Richard received a letter from Thach indicating some success with magnetometers. This closed out this brief episode in Richard's career.

When the BCD paper appeared it refuted the Weiss-DeMarco experiment. The new results left some disagreement in comparing the electron distribution calculated for isolated atoms with those measured on metals. But, more importantly, it showed that such measurements were fraught with difficulty and it would probably take a lot more effort to pin down the fine details of electron behavior in metals. Hence, Batterman had come along at the right time lest the theoreticians continue to soar off into space to their planet of idle speculation!

What went wrong with the Weiss-DeMarco experiment? The two had correctly recognized that extinction would be a problem and had made their measurements at several wavelengths to correct for this effect. So they went back to the 'drawing board'. By a detailed examination of many crystals over a variety of wavelengths they concluded a study that demonstrated the principal fault in making the extinction correction. It couldn't be done by merely varying the wavelength. Weiss and DeMarco had tried to correct for the effect but had been done in by the fickle finger of fate.

With that, Richard took off for Imperial College London for a year. The Army had started its own Rockefeller Awards

Program and Richard was first in line. He asked himself, "Had the entire experimental effort on electron distribution been a virtual waste of time?" He would have at least a year to think about it.

WATERTOWN 3–1961

In 1959 Richard was further haunted about interdisciplinary questions and decided to try an experiment. What if experts in various disciplines met over dinner once a month and explored overlapping interests? He had no idea what might emerge, being unaware of any parallel effort, yet he felt certain it would be fun. Most people will try anything once.

It required a reasonable effort to find the right sort of experts, those who exuded friendship and confidence and were lured to the concept. With a dozen victims enticed to join, the first meeting was held in a private dining room at the Boston Club where a psychiatrist, a doctor, a dramatist, an historian, and a physicist dined and gossiped. What should they talk about? No one knew. "OK," suggested Richard who had prepared some notes, "Let's try out a few topics for size and zero in on one at our next meeting, namely,"

1. Is recognition of his fellowman the sole driving force in man's creative ability or is there something innate?

That question could keep a group going ad infinitum.

2. Is science moving away from other fields, i.e. are scientists unfair to non-scientists?

There's a hint of scientific arrogance in that query, particularly since they were better paid at universities.

3. Should the direction of achievement in various fields be regulated so as to reduce potential hazards?

That sounds like Big Brother but one with a conscience.

4. How are cultural values influenced by progress in various fields?

That needs a definition for culture.

5. What produces fads in various fields?

How does one recognize a fad? It's only easy when you sell a million hoola hoops.

That the group agreed to meet again demonstrated some measure of success. For one thing it indicated that various disciplines could communicate without incurring indigestion. It was also fair to say that Richard's five questions encompassed all the disciplines.

Waiting in the wings for future meetings were a Professor of Comparative Religion, a Professor of Music, a Professor of Political Science, a Biophysicist, an investment counselor, and a sociologist.

The next session included the doctor, the Professor of Comparative Religion, the dramatist, the historian, the psychiatrist, and the sociologist, with Richard taking notes and taping the proceedings. At this meeting the group focused on the specific subject that would occupy it for the next few years until Richard left for England and Alan Levitt for Hollywood. The subject was communication.

The group agreed that communication required a sender and a receiver between which an idea or emotion is imparted. Since the group did not include a prizefighter sending physical blows, pain was not to be included with the transmitted emotions, only thoughts. But assuming an idea or emotion is clear, effective communication still required that the idea or emotion be received with little distortion. The group emphasized that definite relationships were necessary between the sender and the receiver for communication to be effective, i.e.:

a. There must be a need for such transfer.
b. A certain amount of trust should exist between the two.
c. The receiver should have some prior training to recognize the idea or emotion.
d. Both sender and receiver must be prepared to work hard in order to make the process effective.
e. A receiver's rejection of an avant-garde idea may be due to it posing a threat to the identity or security of the receiver.

After engaging in a dialogue between sender and receiver it was important to recognize how the two are changed as a result of effective communication, what the obstacles were to effective communication, and what the purpose was for the communication.

At this point the attendees developed a sharper focus by concentrating on communication within and between disciplines.

At the next meeting the group agreed on specific questions to be individually addressed by each member at future meetings.

1. *What are the specific ideas or emotions you are attempting to communicate?*
2. *How do you recognize successful communication and how are you and your audience changed as a result?*
3. *What are the specific obstacles to successful communication in your field?*

It was also recognized that one universal obstacle to communication was the lack of awareness on the sender's part of the change desired in the receiver. Sometimes the sender's attempts at communication are motivated by the subconscious need to ensure the sender's own sense of security, i.e. the receiver is merely employed as a mirror for the sender's ideas.

In February 1960 Richard stepped up to become the first batter. He gave the following answers to the three questions:

1. A scientist communicates almost exclusively with his fellow scientists via articles in journals, lectures, papers delivered at scientific meetings, books and informal discussions. He either communicates experimental observations or attempts to create mathematical order out of the experimental results. The extent to which a formula or equation explains or predicts experimental observations is the extent of his understanding. Should a formula or equation be capable of predicting all experimental observations in nature, then the scientist as we know him would become redundant.

2. If a scientist's ideas are understood by his colleagues, communication has been successful. While scientists are human and tend to reject new ideas, science is least encumbered by this problem. New ideas can be tested for

truthfulness and both scientist and his audience gain more knowledge about the physical world.

3. The main obstacle arises in trying to communicate an untruth since experts will force the scientist to defend his position. The more truthful the observation the easier to communicate. In addition to conveying the truth, successful communication demands a clear exposition of the facts by the scientist, although this is secondary to being correct. Thankfully, nature is always at hand as the final arbiter in any scientific dispute.

This report of Richard's was rather straightforward and came as no surprise to the non-scientists. Still, it represented a beginning.

Next up to bat a month later was the psychiatrist who answered the three questions thusly:

1. The clinician must establish a feeling of security in his patient to encourage communication. The clinician induces the patient to re-experience the old troubling situations in a new environment and encourages him to discover avenues to circumvent his problem.

In communicating with his peers the clinician relates his experiences with patients through journals, books, technical meetings, etc.

2. Success with a patient induces communication with the clinician (diagnostic) or to himself (therapeutic) enabling the patient to overcome former obstacles and improve his behavior. Success is gauged ex post facto by analysis of accomplishments or failures.

3. The obstacles include:
   a. The clinician personalizes his interaction with the patient and arouses his own emotion. Feeling uncomfortable he tends to isolate himself from the patient.
   b. The clinician overinterprets the problem beyond what is necessary.

    c. The clinician may worry that the patient will tell
       the public he is getting 'sicker' and jeopardize his
       reputation.
    d. Patients who are forced into psychiatry, i.e.
       prisoners or children, require the clinician to create
       a receptive atmosphere.

It was pointed out that most clinicians undergo analysis themselves and that clinicians are divided into organicists for physiological disturbances or dynamicists for environmental disturbances.

Richard felt the curtain lifted for him in the mysterious world of psychiatry. Furthermore this underscored how much simpler it was dealing with electrons than people. The group moved on to the investment field, notably mutual funds. The group's expert wrote sales literature to encourage investors to recognize that:

1. (a) Common stocks were good long-term investments.
   (b) They follow inflation.
   (c) Mutual Funds are a simple way to own common stock.
   (d) Diversity reduces risk.
   (e) The funds are managed by professionals.
2. Success is measured in increased sales, favorable comments about brochures. The audience is changed by developing increased security in mutual funds (as long as income is assured).
3. Obstacles to communication:
   (a) Public's ignorance about buying stock.
   (b) Lack of imagination by the public.
   (c) Stringent SEC control as to the wording of sales literature.

There were hardly any surprises in these answers. To Richard the stock market revealed an element of 'refined' gambling. Thirty years later the mutual fund business grew by an order of magnitude even though prospectus literature continued to read as obtusely as insurance policies. Gambling through lotteries,

and through Wall Street investments had become a way of life —
man lived on hope.

A more subtle discipline, radiology, was up next and that
offered a few surprises:

1. A radiologist communicates primarily with physicians who
send him referrals. He must describe in words the content
of a picture, one which frequently reveals only the faintest
hint as to the problem. It's like identifying the shade of
green on a distant hill from a black and white photo taken
on a foggy day. Based on experience the radiologist must
estimate the presence or absence of a suspected condition
from the x-ray. The language used depends on the
radiologist's knowledge of the attending physician's
respect for the difficulties in radiology.

2. Successful communication occurs when the attending
physician neither overemphasizes nor underemphasizes the
radiologist's report vis a vis all other symptoms.

Richard, himself, employed x-rays by scattering them from
materials while the radiologist sent the x-rays through the patient
and determined how much they were absorbed. Same x-rays,
different technique.

The obstacles included:

(a) Physicians may withhold vital information that would help
the radiologist in his diagnosis.

(b) Physicians may insist on a positive ID, rarely possible by a
radiologist.

(c) The patient or his doctor does not make previous x-rays
available for comparison. Differences in x-ray negatives are
easier to distinguish.

(d) Physicians may have preconceived ideas about the patient's
condition and may 'vie for position' with the radiologist.

The Radiologist informed the group that it was a good
idea to keep old x-rays although as Benjamin Franklin advised in
Poor Richard's Almanac, "God heals and the doctor takes the fee."

At last Richard hit the jackpot in identifying a communication problem. The radiologist had sparse information,

the attending physician may have operated as a result, and the radiologist may have had to guess how well the physician was capable of diagnosing a condition. At the time of this meeting (June 1960) the laser was just discovered. Thirty years later the endoscope, laser surgery, and other forms of diagnostic tools like magnetic resonance imaging and ultrasonic imaging would go far in clarifying the radiologist's position. But in 1960 he clearly had a heavy burden and a communication problem!

Up to the plate at the next meeting stepped Professor Black, the historian, who provided the following:

1. History has been defined as the study of everything that has happened to human beings. While an historian must devote considerable effort to uncovering and documenting past events, gaps in our knowledge force historians to interpretation. This involves:
   a. Elucidating the manner in which historical facts influenced the subsequent course of history.
   b. Establishing contemporary sets of values and patterns of behavior on the basis of the past. This is the major cause of controversy and delimits a scientific approach to history. As someone said — an historian is a guy who wasn't there.
2. Successful communication is achieved when the audience adopts a pattern of behavior consistent with these values. The influential historian becomes the creator, guardian, and revisionist for traditions.
3. The obstacles are myth, prejudice, and dull historical authors.

A number of ancillary points were underscored concerning historians in their approach to the subject:

(a) They expostulate reasons for the cyclical and continual rise and fall of civilization.
(b) They attempt to establish rules for systematic historical change.
(c) They uncover and emphasize neglected factors in historical interpretation,

(d) They contend interpretation is unnecessary — just the facts, ma'am.
(e) They deny that objectivity is possible. History is continuously reinterpreted in the shadow of contemporary influences.
(f) They admit to the limitations with objectivity but don't concur with generalized hypotheses.

History has been an academic discipline for 200 years and it takes 20 years for an historical find to end up in the textbooks. But even if all the facts are known historians will still disagree.

Most students hate history because it concentrates on the memorization of dates, probably the most certain aspect of history. But presentation of history was beginning to alter in Richard's mind. Richard's later passion in restoring and operating an historic tavern was based on reviving a living history.

The artist expressed the role in society that his artistic creations fulfilled:

1. An artist attempts to arouse in his audience an insight into his own emotional and intellectual attitudes toward his environment.
2. The artist's conception and execution are not oriented to his audience although 18th century portrait painting had to please the sitter and pre-Renaissance artists were subject to church control. Only after completion of his work does the artist seek audience response. In the former stage, success with his technique (line, shape, color, etc.) intimately expresses his emotional attitude towards his environment. He is changed by sensing a power to express the gamut of his emotional attitudes through his work. In the latter stage successful communication is recognized by the tone of voice of the non-artist and by the critique of his fellow artists. The audience sees new truths, or a re-emphasis of old truths or simply a new means of expressing a truth.
3. Obstacles include: the artist's lack of technique, sensitivity, perception, or intelligence; the rejection of an idea as a

threat to the viewer's security; the audience's lack of education, culture, historical background, etc.

Notes:

(a) The eye and brain translate technique (line, color, shape, composition) into a specific emotion. The artist responds to the world visually.

(b) When artists congregate they discuss technique, emotional responses are left tacit.

(c) Rembrandt has been the universally successful artist.

Richard found that scientists tended to prefer music more than art for their cultural stimulation since their scientific objectivity related more readily to the ordered regularity of the diatonic scale. During this period Richard socialized with a number of well-known New England artists most of whom viewed him with suspicion. Only by extended exposure to tens of thousands of modern paintings did Richard feel a familiarity and emotional understanding to abstract painting. It was a culture that defied verbal description although there were many critics that made the attempt. To the novitiate the impulse to identify abstract art with recognizable real world entities obstructed the path toward entering the artists inner being.

The only comment Richard could make about abstract art was that it was possible to recognize competent technique but emotional content depended on individual response. Unlike the scientist who anchors his subject to the laws of nature, the artist is under no such restriction.

With the huge sums spent at auctions for certain artists the investment counselor had been drawn into the field, diverting the public's attention away from the cultural aspect. Picasso has criticized this aspect of the art market with these remarks:

"I'm no pessimist. I don't loathe art because I couldn't live without devoting all my time to it. I LOVE IT AS THE ONLY END OF MY LIFE. Everything I do connected with it gives me intense pleasure. But still, I don't see why the whole world should be taken up with art, demand its credentials, and on that subject give free rein to its own stupidity. Museums are just a lot of lies, and the people who make art their business are mostly imposters."

Alan Levitt then gave the group his thoughts about drama:

1. Drama originated in the secularization of the religious ritual and through the years has been a visual re-enactment of the cycle — life, suffering, death, judgment, and resurrection. Drama attempts to communicate to its audience the emotional and intellectual conflicts during this cycle and permits the audience to share the experiences of the protagonist without, in fact, suffering his judgment.

2. Successful communication can be recognized during a performance by a close inspection of the collective behavior of the audience (breathing rate, body position, etc.) as it coincides to the emotional experience of the protagonist at the time. Following the performance a feeling of relief coupled with a sense of purpose and introspective clarification should be felt by the audience from the experiences shared by the hero.

During the successful moments of a drama there is a collective awareness amongst the performers and audience that both are sharing a common experience. This serves to enrich the purposefulness of the moment for both audience and performer. Successful communication can be recognized by a comparison of the attitudes and emotions of the audience with that intended by the playwright.

3. The obstacles are:
   (a) Bad playwriting — it is difficult to find an action for the actor that justifies the words in the script.
   (b) Bad acting — it becomes difficult to find an audience.
   (c) The audience lacks the historical, cultural, and ethical background to understand the attitude and emotional responses of the characters to their environment. This is especially true in translating drama between ethnic groups or at different historical periods.

Notes: The collective presence of an audience can be felt during a performance. The group feels this awareness as a sense of 'oneness.'

Ibsen, Shaw, and Brecht employed the conventional cycle of life, suffering, etc. for social propaganda.

The final meeting of the group was privileged to have Professor Amiya Chakravarty talk of the communication problems in religion. His answers were:

1. Religion is a living guide to life in all its menial aspects. Its good is to make a satisfactory experience of one's life, to accept people as they are, and to develop a conscience for the plight of man.

A religious man should be aware of the physiology, history, psychology, legal systems, etc. of mankind rather than delimiting his view to the non-biological institutional aspects. A living religion is one that does not impose illogical dogma and tradition through rigid adherence to senseless customs, particularly since the practical reasons for their origin may no longer exist. Such a religion is devoid of supernatural events and the Divine Being becomes a device of the mind. The religious ritual takes on a poetic aspect and one can enter and leave it as one pleases.

2. Successful communication is recognized through complete acceptance by others without a conscious or subconscious desire to alter your beliefs. A feeling of security and faith develops between individuals.

3. Obstacles include established institutions, intolerance, fetishes, and artificial barriers to new ideas.

Notes:

(a) Christ and Buddha closely followed these principles.

(b) The resurrection of Christ at Easter was made to coincide with the vernal equinox.

(c) A 'Supreme Being' can be understood in the Einsteinian 'sense of design' in the universe.

Having encompassed a wide enough range of disciplines to search for a common denominator Richard tried to find some universal truth. Thirty years later, in the last decade of the 20th

century, the very element silicon that Richard had considered useless when he first arrived at Watertown had created a communication revolution with its ability to store and process information in a space many orders of magnitude smaller. The fax machine, the modem, and the videotape had placed at man's fingertip the knowledge and lore of the world. The various disciplines that met with Richard in 1960 and discussed their communication problems still existed as the world poised to enter the new century. Mankind faced increased longevity from better control of disease although natural disasters and a plague like AIDS indicated how vulnerable he was to life-threatening situations. The dinosaurs disappeared — it could happen to man. Such a cataclysmic extinction would scatter into outer space the sonnets of Shakespeare, the theories of Einstein, the music of Bach and Mozart, and the teachings of Christ and produce but an insignificant increase in the entropy of the universe. To this tragedy God might only cluck His tongue and wait for civilization to start over again elsewhere.

IMPERIAL COLLEGE LONDON: 1962–1963

Imperial College in London's South Kensington district is situated a few miles west of Trafalgar Square and but a quarter mile south of Hyde Park. Its roots can be traced back to the early nineteenth century when the American scientist Benjamin Thompson, former spy and Loyalist, had returned to London after a tour of service with the King of Bavaria. Now holding the title Count Rumford, he proposed an institution to display the latest in technological achievement. Art museums were commonplace but not science museums. From this idea emerged the Royal Institution on Albermarle Street in London, dedicated to displaying and educating the public in the latest scientific accomplishments. Its Friday lectures were attended by the elite who would be entertained with demonstrations of nature's mysteries. The Institution's programs were so well attended that Albermarle Street became the first one-way street in London, a necessity to handle the increased coach traffic.

Several decades later Prince Albert, very much in favor of the Royal Institution's efforts, proposed the Great Exhibition. This suggestion was avidly received by prominent businessmen and in 1851 it successfully displayed the world's technology at the Crystal Palace in Hyde Park. In fact, the Great Exhibition became the first World's Fair. It also led to the America's Cup when an international sailing competition celebrated the event by circling the Isle of Wight. It was won by the sailing vessel 'America' and the winning ship was presented the famous Cup by Queen Victoria. With funds left over from the financially successful Great Exhibition Prince Albert proposed the creation of the Imperial Institute south of Hyde Park to perpetuate the enhanced interest in science. The original concept was expanded to include several museums and later, Imperial College.

Richard arrived in 1962 and was welcomed into the Mathematics Department by Peter Wohlfarth and Harry Jones, two of the school's leading lights in theoretical solid-state physics. Maths was housed in the archaic five-story Huxley Building on Exhibition Road, probably one of the original structures from the Prince Albert period. It featured a rickety, open-latticed lift, and a stone circular staircase worn concave over the years by those who gave up waiting for the lift. As much as Cambridge reeked of its Gothic inheritance so did the South Kensington complex exude Victoriana.

Across Exhibition Road from the Huxley Building were the Natural History Museum, the Geological Museum, and the Science Museum while just to the south and abutting Huxley was the Victoria and Albert Museum (V&A). Busloads of children were disgorged daily for their field trips into the museums. A tunnel provided pedestrian access under busy Cromwell Road to the South Kensington Underground Station a quarter mile distant. Equally distant to the north was Hyde Park with its architecturally bizarre Albert Memorial and the almost sacred Royal Albert Hall, a large auditorium with good acoustics that became the venue for the popularized Henry Wood summer concert series. The Hall had a posh atmosphere, the volunteer ushers being selected from the respectable levels of society, with a long waiting list for replacements.

Richard lived at Vicarage Gate, a more desirable part of London just to the west of Kensington Gardens and Kensington Palace. The Palace was home to Princess Margaret and her collection of dogs which could be seen twice a day being chauffeured the 100 yards to the park entrance.

The one-mile walk from the Huxley Building to #12 Vicarage Gate crossed the well-tended Kensington Gardens and took Richard south of the Palace and across Palace Green, a gas lit road containing over a dozen Embassies including the infamous Russian Embassy at #13.

Every day Richard walked past the blue-plaqued home that formerly housed the famous physicist James Clerk Maxwell when he was a member of the King's College London Physics Department and would either walk or take a carriage for his 2½ mile journey to work. Little did Richard realize that many years later he would temporarily occupy the office of the Maxwell Professor at King's College while the school sought a replacement for the retired chair holder Cyril Domb.

The surroundings appeared pleasant and interesting as Richard idly wandered afield from Vicarage Gate but he was still anxious to do some physics. As at Cavendish in 1956 he did not have access to experimental equipment so he had to work on problems that could be solved with pencil and paper.

Richard found it is sometimes useful to be forced to exercise scientific logic. He took up the problem of the two forms of iron, magnetic alpha iron and non-magnetic gamma iron. Was gibt es? The specific heat of iron clearly showed that something unusual would happen to gamma iron if one could stabilize it at low temperatures. He walked the paths of Kensington Gardens seeking inspiration and even trod Baker Street looking for 221B, the address Conan Doyle used for Sherlock Holmes' digs. Damn! There was no such number and no plaque to show reverence for the famous detective. Many a tourist made the mistake of walking past the spot where 221B should be, feeling foolish and believing everyone was watching him and having a good chuckle over their dilemma.

The inkling of a solution to this enigma over iron came to Richard during a conversation with a few other physicists while enjoying a brew at a pub. At first blush it was a crazy idea. Maybe gamma iron is different at low temperatures than it is at high temperatures, i.e. it slowly changes identity on the way up in temperature. Richard lived with this ridiculous thought for a few days, then sat down in his Victorian office and undertook some calculations to explore this untrodden path into a jungle of speculation. When James Clerk Maxwell developed the revolutionary Maxwell's Equations his mind often wandered afield and he sometimes resorted to non-physics images to keep the 'ball rolling'. Richard never analyzed his own thinking process but it was probably true that successful physicists were uninhibited. Others may worry that they'll be ridiculed if they espouse an idea easily shot down. No matter, thought Richard, pick yourself up and try again. Nature yields up her secrets reluctantly.

Reduced to simple terms gamma iron was postulated to come in two varieties g_1 and g_2 with different mass densities. There was a small energy difference between the two and as one heated up a sample one would gradually switch the one state to the other, atom by atom. Richard went on to explain the so-called invar effect, i.e. an alloy of iron and nickel or iron and platinum that would go through a region where it actually shrank with increasing temperature because the g_1 and g_2 atoms were of different sizes.

Once the door was opened to this zany hunch it not only explained the specific heat of gamma iron but the effect was also discovered in other elements like manganese and chromium. It spawned a new area of investigation for the 3d elements and gained a small following. Generally, new ideas were a dime a dozen but an idea that manages to be useful for thirty years was well worth the pangs of birth suffered by the father.

The two gamma states of iron was not Richard's sole discovery during his Imperial College days. Portobello Road, a fifteen-minute walk from Vicarage Gate, resurrected English history through the antique artifacts that were peddled, particularly on a Saturday. It probably accounted for Richard's developing a keen interest in history. Continued examination of antiques peaked Richard's curiosity about the people who designed and fabricated them and the social interactions of those peoples. Furthermore, 1962 was a good time to buy antiques since prices were depressed and Americans were relatively affluent in comparison to English levels of income.

Richard purchased three grandfather clocks (circa 1790) for a song. They were so cheap they had to be genuine — no one could fabricate them even at 100 times the sale price. There was some interesting physics here:

Huygens, the Dutch scientist invented the pendulum clock about 1660 when he discovered that the period of a pendulum was constant as long as the displacement from the vertical was small (less than a few degrees). Hence this provided the steady beat needed for good timekeeping. The period only varied if the length of the pendulum changed, as with temperature, but this could be easily adjusted. The long case was fabricated to hide the pendulum and the terms long case clock, coffin clock, and pendulum clock became synonymous. But about 1850 a songwriter published the catchy tune called 'My Grandfather's Clock' merely referring to the clock that belonged to his grandfather. The song caught on and long case clocks were thereafter referred to as grandfather clocks.

The demand for such clocks increased in America to the point where the brass technology for the gears could not keep up. Wooden gears began to be utilized but they were not as reliable for timekeeping. Long case clocks could not be used by mariners since one required a level floor and it wasn't until the end of the 18th century that a reliable spring-loaded clock was fabricated. This was necessary at sea to be able to determine longitude relative to Greenwich with but three pieces of information, the time and date, the sun's elevation above the horizon, and the navigational tables of Bowditch.

Richard also became fascinated with pewter, an alloy of tin that was used in pre-18th century pubs. Why did pewter look so dull and how does one put a shine on it? More physics. Tin is actually a silvery metal that does not oxidize too readily but once it does, the gray oxide is tough and not easily removed by polishing. Richard noticed that the color of oxide-free tin was remarkably close to that of silver. Furthermore, if one examined a selection of 'silvery' metals one found distinct hues. This had to be explicable in terms of the way sunlight is reflected or absorbed by the electrons at different wavelengths. Richard took up this subject when he wrote his first book on Solid State Physics. Thank you Portobello Road.

Gold, silver, platinum, and copper. Why were they so very different in color and why did they tarnish so differently? Why did woods have such varied appearances and physical properties? Why were gems like opals, diamonds, sapphires, and rubies so different in hardness and color? These and other questions aroused Richard's interest. Thank you again, Portobello Road. The last time Richard wandered down the streets of this part of London he guessed there were at least 1000 merchants selling from pushcarts and stalls. The history of the ordinary man for the last few centuries was bared through these artifacts.

Along the way Richard was impressed with how little dishonesty he found amongst the antique dealers of Portobello Road. As a group scientists were also fairly honest. Perhaps this was one reason Richard liked wandering through the area and chatting with the merchants.

Paintings captured Richard's fancy, particularly 18th and 19th century landscapes and seascapes. They were often well executed and inexpensive. At one point Richard was offered an impressive classical revival painting for £80. It was painted in 1820 by someone listed in Benezit, the handbook for recognized painters, but needed a cleaning. Richard turned it down — it measured 10' × 8'!

With electron distributions only marginally invading his thoughts, Richard had to find his science wherever it popped up. At a small conference in London Richard met Harold Urey, Nobel Laureate of 1934 who discovered heavy water. He had given an after-dinner talk on the abundance of the elements in the universe when a very seedy-looking member of the audience got up during the question period,

"Professor Urey. According to your numbers on the abundance of chemicals in the universe there is only one molecule of iron oxide in the solar system."

Everyone laughed since Urey was only making crude estimates of abundances. Urey was ready to go on to the next question when the man added a bon mot,

"Just a moment! My question, Professor, is, 'Where is that molecule?' "

Over coffee Urey confided to Wohlfarth and Weiss that recent seismic data showed that the iron-nickel core of the earth had a solid center. With this, Weiss raised the possibility that it was this solid core that accounted for the earth's weak magnetic field. The origin of the earth's magnetism was uncertain at that time although the best guess was that it was due to circulating currents in the liquid part of the earth's core. The next day Richard sat in his office in the Huxley Building and did some simple calculations to follow up on his idea. He concocted a nebulous argument to support his cause and it was published in Nature. As nearly as he can remember virtually no one paid any attention to the paper. He spoke to P.M.S. Blackett, Vice Chancellor at Imperial College and 1948 Nobel Laureate for work in nuclear physics. After the war Blackett went on to consider questions about the magnetic field of the earth, the sun, and stars but he showed no interest in Richard's work. You win some — you lose many more in theoretical physics.

Richard read in the papers that the retired Winston Churchill was living in Hyde Park Gate, a short street south of Hyde Park. Richard wandered down the road one afternoon feeling as ridiculous as the time he looked for 221B Baker Street. Foremost in Richard's mind was the Englishman's desire for privacy. He walked rapidly to the end of the road, a cul de sac, and retraced his steps looking for Churchill's cottage. And there he was — standing in a bay window, the cigar in visible evidence to support the distinct visage and slightly obese body that completed the picture. The great man looked directly at Richard who was standing alone but Richard was too embarrassed to stare back or wave. If only the cigar-smoking Richard had had the courage to light one up, wave, give the old V for victory sign, and then move on. But, when in London don't do as the Americans do.

One day Professor Wohlfarth indicated to Richard that Prince Philip was visiting Imperial College and that he'd be passing along the corridor. Richard waited and, sure enough, the Duke of Edinburgh turned the corner and came directly toward Richard. Thinking that Richard was next on his list of Academics to visit, the Prince extended his hand to shake. Poor Richard was too baffled to know what to do and he stood there like a statue in Madame Tussaud's until the appropriate person appeared and led the Prince away. Richard told the story to his colleagues, ending with,

"Anyone can shake the Prince's hand but how many have the courage to refuse to shake his hand?"

A few years prior to Richard's sabbatical at Imperial College he had approached Harvey Brooks about writing a book on physics but aimed at metallurgists. Harvey liked the idea and recommended it to Robert Maxwell. Richard signed a contract with Pergamon Press to produce a book entitled *SOLID STATE PHYSICS FOR METALLURGISTS*. The concept was new and required a dedicated effort to avoid a book that looked or sounded like a physics text, i.e. Richard sought to eliminate tedious mathematical derivations that would cause metallurgists to lose their way. Richard's philosophy was injected into an opening tale that appeared to have biblical overtones:

> *On Mount Gerizim in Samaria dwelt the "miracle maker" called Ramuth Gilead. There came to him the first born of old and ailing King Cyrus, young Bezer, who begged for infinite wisdom to lead his people.*
>
> *"Infinite wisdom I cannot grant, but I bestow upon you all the wisdom of mankind. Go forth and use your knowledge wisely."*
>
> *And so Bezer returned to his people and he was crowned King upon the death of his father.*
>
> *Many years passed and his people prospered and they blessed Ramuth Gilead for the wisdom of their King. But in the fourteenth year of his reign King Bezer again made the journey to Mount Gerizim, and he came before the "miracle maker" saying,*
>
> *"In my sweet youth I knew not of the Heavens and the Earth and of the sciences and so I did sleep well. But I have drunk deeply of the wisdom you*

have bestowed upon me and have looked into the
unknown. The many questions I now ask myself
keep me from my sleep."
And answered Ramuth Gilead,
"That is the price you must pay for wisdom."
And King Bezer, who was now full grown of beard,
thought deeply and asked,
"How long must this endure?"
And Ramuth Gilead answered,
"Until you know nothing."

Several years later Richard received a Russian book in the mail. Not reading Russian he examined the book and thought he noted something familiar. It was, in fact, his own book translated into Russian and the familiar aspect was the figures. Neither Richard nor the publisher had granted permission for this example of Iron Curtain plagiarism. A notable absence, though, was the exclusion of the 'biblical story', which even the Russians should have realized was fabricated by the author!

When Addison Wesley inspected page proofs they showed them to Professor Morris Cohen, Chairman of the Materials Science Department at MIT and future recipient of the Japanese 'Nobel Prize'. Cohen approved of Richard's effort to introduce the subject through the experimental approach. He felt this helped the metallurgist immerse himself in solid-state physics in a realistic way and kindly provided a Foreword for the book.

Was Richard's approach a good idea? It was certainly a learning experience for Richard and it received good reviews but it did not sell in the conservative community called Academia. Twenty-seven years later, after the book was out of print and the copyright had reverted to the author, the book was republished at five times the original price by another company. It still hasn't sold.

Science is taught at universities in a staid, formal manner. The lecturer's notes end up in the students' notebook without passing through anyone's brain. Einstein learned a lot of physics by thinking about patent applications when he was a lowly clerk in the Swiss patent office. If a physicist wished to get close to God let him become an experimentalist, thought Richard. If he wished to create his own god let him become a theoretician. Experimental physics was emphasized in Richard's book.

It was at this time that Richard conceived the idea that all theoretical papers should contain error bars. Experimentalists always quote the estimated errors in their results based on a personal and honest assessment of how badly things can go wrong. The error bars might be underestimated but at least an effort was made to determine what might have gone amiss in the experiment.

Not so with the theoretician who cannot attack a problem without simplification. If the theoretician would spend time in asking himself how his simplifying assumptions affect his answer he might be able to make an assessment of his error bars. Alas! They refuse to do this!

John Williams, the now-famous classical guitarist from Australia, gave his first London concert in 1962. He was young and unknown but Richard was enchanted with his playing. He asked Williams if he would give him lessons but was turned down — he did not teach privately. However, he did recommend to Richard a guitar teacher in Chelsea, London. Hector (Bill) Quine turned out to be a patient teacher and Richard developed a passionate interest in the classical guitar.

Over the years Richard met most of the famous players like Segovia, Bream, Yepes, Liona Boyd, etc. The physics of the guitar captured Richard's interest and this was shared by Quine who also constructed a few guitars. But Richard eventually concluded that physics could not help the luthier.

The modern classical guitar emitted a distinct sound that had evolved over the centuries from the empirical efforts of many; the luthier (artisan), the composers, the artist and his playing technique, the woods and the string material, and the dimensions of the instrument. Alter any of these and one did not achieve the sound the audiences had learned to recognize and appreciate. The psychology by which the acme of achievement in all these areas passed through the human auditory system and was correlated by the brain to release a sense of well being in the listener was too mysterious to comprehend.

The ear and brain, trained over years in the nuances of music, had acquired some remarkable degrees of perception. It could recognize in less than a second a plucked guitar from a variety of other plucked instruments, it could store thousands of complete scores of music each involving as many as 10,000 notes and reproduce them in proper order, it could differentiate successive notes spaced only 1/100 second apart, and it could distinguish two plucked notes even though a harmonic analysis by a physicist showed them to be indistinguishable. In short, the subtleties of this complex human system defied scientific unraveling. This did not imply that paranormal phenomena were actively at play. The ear and brain was simply a remarkably efficient instrument with billions of functioning components that one could not reproduce in the laboratory.

Yet the physicist has tried to improve the guitar. Unfortunately one had to define what one meant by improvement. The guitar had a quiet voice and there were times when an increase in decibels would be desirable as in a guitar concerto when thirty other instruments were being sounded. Whether it was possible to change something in the guitar's structure or playing technique to increase the loudness without altering the tonal quality was moot. Today that is invariably achieved through amplification employing a well-designed electronic system of pickup and loudspeakers that is relatively free of distortion.

About thirty years ago Richard attended a scientific conference of physicists in Canada with an after-dinner talk by a physics professor who recounted his own experience with guitars. He indicated that his first examination into the construction of a guitar convinced him that "it was all wrong." He redesigned the instrument even though he was not a player and he delivered some rationale at the conference for the changes. Richard was unable to comprehend the speaker's logic. Anyhow, the upshot of the story was that Liona Boyd, a Canadian guitarist of considerable talent

Age 6

Teenager

New ensign   Outside main gate Annapolis
                                    Chapel behind

New PhD
New York University PhD, 1951

Serenading daughter
Note parakeet on wrist

J-STRUCTURE STUDIES
NEUTRON DIFFRACTION

WATERTOWN ARSENAL LABORATORY   WTN. 639-12,148

Demonstrating neutron diffraction at Brookhaven, 1953

Cover drawing for Richard's book, *A Brief History of Light*

Richard congratulated for his Rockefeller Award

GREAT HALL
NATIONAL GALLERY OF VICTORIA
22ND AUGUST, 1974

Banquet hall in Melbourne

Alexander Solzhenitsyn strolls in downtown
Montpelier, Vt.   (UPI photo)

Alexander Solzhenitsyn in Vermont
— Richard's lookalike

CAVENDISH LABORATORY
CAMBRIDGE

COMMEMORATION DINNER
to mark
The Hundredth Anniversary of the Birth of
SIR J. J. THOMSON

Saturday, 15th December
1956

TOASTS

THE QUEEN
—

TO THE MEMORY OF Sir J. J. THOMSON
The Master of Trinity
—

THE CAVENDISH LABORATORY
Proposed by
SIR JOHN COCKROFT

The Cavendish Professor of Experimental
Physics will reply

MENU

Amontillado	Consommé Julienne
	—
	Filet de Sole Meunière aux Crevettes
	—
Beaujolais 1952	Caneton Rôti
Zeltinger 1949	Petits Pois
	Pommes de Terre Rissolées
	—
	Coupe aux Fruits
	—
	Croûte Baron
	—
Croft 1945	Dessert
Sercial	—
	Café

Cavendish Laboratory commemorated the 100th anniversary of the birth of J. J. Thomson, discoverer of the electron

31 January 1964

Dr. R. J. Weiss
U. S. Army Materials Research Agnecy
Watertown, Massachusetts

Dear Dr. Weiss:

I have read with great interest your report of your
activities in England under the Secretary of the Army's
Research and Study Fellowship. The trip report is a
fascinating document. It is most gratifying to see the use
you were able to make of the Fellowship, although I don't
see how it is possible for one man to do so much in so
short a time. I wonder how you were able to complete six
papers, deliver a course of lectures on the 'Nucleous as a
Tool in Solid State Studies', and visit as many people and
places as you did, -- and still find time for eating and
sleeping.

I am grateful for the example you have given your
associates in the Army Research and Development establishment of the potential of the Secretary of the Army's
Research and Study Fellowships, and wish you all success
on your return to the laboratory.

Yours very sincerely,

Charles L. Poor
Deputy Assistant Secretary of the Army (R&D)

Letter from Asst. Secretary of Army

30 April 1991

Mr. Richard J. Weiss
4 Lawson Street
Avon, Massachusetts   02322

Dear Mr. Weiss,

It's always good to hear from a CCNY alumnus, especially when they've been as successful as you obviously have. It just proves what I've always believed -- CCNY afforded us as good an education as any school in the land.

And I'm very grateful for your kind words. Let me congratulate you as well. Physicist, tavernkeeper, trustee -- I don't know which I envy you the most, but like Benjamin Franklin, I'd probably have to opt for the tavern if I were forced to choose. And for the 30-year career with the Army as a civilian scientist, I can only say well done. I know you enjoyed that experience.

Thanks again for your letter and best wishes,

Sincerely,

COLIN L. POWELL
Chairman
Joint Chiefs of Staff

Letter from Colin Powell

With nobelist Cliff Shull at Blanchard's Tavern

Prof Malcolm Cooper (left) and Richard at the Avon Strawberry festival, July 1976. Prof Cooper had organized the 15th Sagamore conference (13–18 August 2006) in the UK.

gave a concert playing the new instrument. Unfortunately she did not play her own guitar for audience comparison.

Of course the new guitar sounded very pleasant played by Miss Boyd but there was nothing Richard could conclude about the tonal quality of the new guitar. So after the concert Richard managed to speak privately to Miss Boyd.

"What did you think of the new guitar design?" he asked.

"I was afraid you were going to ask that," she countered.

That said it all. She was well paid to play the new guitar but that didn't mean she had to like it or to endorse it. She fell back to her own guitar, fabricated along traditional lines, for her concerts.

The only ingredient in the modern guitar that has benefited from modern technology is the nylon string, a man-made material that is more uniform than nature's sheep intestine (gut) or the silk worm's cocoon (silk), both used before nylon came along.

One of the interesting applications of physics to the guitar was the use of pulsed lasers to photograph the standing wave displacements of the sounding board when the guitar is played. These Chladni patterns were aesthetically interesting but, like accounting for the missing 3d electrons in iron, one can employ all sorts of rationalizations to argue what the patterns should look like in a 'superb guitar'. And no matter what the physicist claims would improve the instrument, someone would have to fabricate such a guitar and this is time consuming. No luthier, unless he was naive, would be taken in by a physicist's claim that he knew how to make a better guitar.

Richard recalled overhearing a spectator who went back stage after a performance by Segovia and asked but one question,

"What make guitar were you playing?"

"It's a guitar!" was Segovia's curt reply.

The implication was clear. The differences between guitars were infinitely smaller than the differences between players.

While physicists can easily intimidate non-physicists with their knowledge of mathematics and complicated physical concepts they must face the fact that music appreciation is much more complex than atomic behavior. The only contribution

Richard, himself, had made to benefit his classical guitar colleagues was to point out the atmospheric conditions that might cause the wood to crack, and even here common sense was sufficient.

One other aspect of guitar playing that had occupied Richard's attention, but to no useful outcome, were the problems associated with fingernails.

Classical guitarists rely on their fingernails to pluck the strings. A plectrum can only pluck one string at a time whereas the independence one achieves with fingernails gives the player much greater versatility. The fingernail had to be strong and not brittle. Break a fingernail and the guitarist cannot function for weeks. Richard often broke fingernails and this led him to seek a strengthening agent.

There were products on the market, essentially for women, from false fingernails to coatings but they were of marginal value. Richard found a doctor in Boston concerned with 'sick' fingernails, i.e. gnarled, unsightly nails that grew irregularly. No remedy was known for this ailment.

One of Richards's colleagues suffered from such disfigured nails on his feet but only the first, third, and fifth toes, the other two toes were normal. What might it be?

An inspection of a healthy nail reveals a graininess along its growth direction, much like the grain in wood. Richard started his research by employing x-ray diffraction on nail samples. Once the word got around about Richard's study he received nail clippings from everyone he met. People proved to be most generous in the interest of science.

Richard discovered nails to be 'crystalline' with many sharp peaks in their x-ray diffraction patterns and Richard found no two nails alike. He realized that nails could be employed in forensics like fingerprints.

After this preliminary study Richard examined both healthy and 'sick' fingernails. Alas! No difference could be identified employing x-rays. Unfortunately Richard's mission at Watertown did not include any such investigation. You couldn't fight the Russians with fingernails. The project had to be abandoned.

Thus, each member of the guitar community has had to learn how to cope with his own fingernail problems. Funny, there are billions of dollars spent on nail cosmetics but almost nothing on research into the basic causes for unhealthy or brittle nails.

In early 1963 one of the London newspapers published a letter to the editor citing the recent adoption of the centigrade scale in England as the obvious cause of the worst winter in over 200 years. The author claimed that when the country had employed the Fahrenheit scale the weather had been consistently warmer. If the letter was a sincere observation of the horrors of the 1962–1963 winter it was certainly not written by a physicist. However if it was a piece of satire, it might have been penned by an engineer frustrated over readjusting his life from Fahrenheit to Centigrade.

The England of 1962–1963 was unprepared for four months of freezing weather and its accompanying snowfall. A London snowplow consisted of two Irishmen and a shovel. While the average winter temperature in England was well above freezing, the infamous winter of 1962–1963 averaged −7F compared to some prior bitter winters such as 1684 (−9) and 1740 (−7.8). Even the dismal postwar winter of 1946–1947 only averaged −5.1. It was not unusual to hear of students at Cambridge University having to crack the ice in their washbasins to rinse their faces.

Richard suffered. Oh, did he suffer! Even at #12 Vicarage Gate in the posh part of the West End Richard tried keeping warm with a gas-fired heater (so the Gas Board countered by lowering the gas pressure) and an electrically heated oil heater (whereupon the Electricity Board lowered the voltage). Nor did the old Huxley Building effectively warm up the mathematicians confined therein. If you drank tea all day you could warm your innards but not your toes. Who could find the inspiration to manage theoretical physics in atmospheric conditions that froze one's ideas.

The effects of sub-freezing temperatures emerged in all aspects of English daily life. The drains in most buildings were external. Everyone knew that sooner or later they would freeze up, the only mystery was when. The London of December 31, 1962, a holiday time when it would ordinarily bustle with New Year's activity, went silent with roads blocked, buses not running, and cars buried in snow. The typical Londoner, by tradition aloof from his neighbors, suddenly discovered who they lived next to during their cooperative efforts to push cars out of the drifts, although at the end of the winter they returned to their contented lives of isolation.

The snow in Kensington Gardens and Hyde Park had thawed briefly but then turned to ice overnight. It was difficult for Richard to find safe footpaths from Vicarage Gate to the Huxley Building. Taking the Underground involved just as much walking with the greater risks of having to cross trafficked roads. The Thames froze but no fairs were held as were depicted in paintings of the 1700's. Everyone talked about the weather. The railways, still relying on steam locomotives, suffered interminable delays in thawing the water and shoveling the ice-encrusted coal. The country ground to a halt.

One of the tragedies to the robust Englishman was the large number of soccer matches that had to be cancelled because the grounds were unplayable. A leading Soccer Club installed electric heating wires in the ground, a satisfactory solution to making the playing field manageable but as one manager declared,

"If the fans can't get to the games what good are the heating wires?"

A group of men who had been making preparations for an Arctic expedition was able to practice igloo fabrication. All one needed was some compact snow and a garden trowel to form the construction blocks. A neighbor to one of the explorers entered his garden to inspect the igloo. He burst out laughing to see his neighbor sitting in the igloo he had just completed and properly attired in a tie.

The first snowfall that winter was on November 17 and the last on March 22. On top of all that miserable weather the electricity workers went on strike, actually work-to-rule. This meant no overtime, no handling emergencies, no anything. No wonder the voltage was reduced! At least Richard benefited from one thing. The beard he had grown was a good insulator.

With the spring thaw Imperial College was rapidly transformed into a new school. Except for the original Imperial Institute tower preserved for sentimental reasons all Victorian structures were razed save the Huxley Building, which was returned to the V&A Museum. The Mathematics Department was scheduled to be the last to occupy new quarters.

Richard prepared his thoughts for his return to Watertown hoping to make some effective incursions into the determination of electron distributions. He considered measurements on silicon and germanium since these were available as perfect crystals. It was believed that a crystal devoid of flaws was the one case where a measurement of the x-ray scattering would yield accurate answers for the electron distribution. He decided he would follow this route. But unbeknownst to Richard an event took place that would have a profound effect on determining electron distributions.

It started just before the winter descended in November 1962 with one of W.H. Taylor's graduate students, John Leake, who decided he would measure the Compton scattering in iron. Richard had visited Cambridge and had some discussion with Taylor and Leake about this most difficult measurement.

Arthur Holly Compton had received the Nobel Prize in 1927 for one of the most elegant measurements of his time. By bombarding a sample of graphite with x-rays of a single energy and analyzing the scattered x-rays he discovered that some of the x-rays lost energy. Compton tried several concepts to explain this loss but finally came up with the simple answer that each x-ray had collided with a single electron and that one could treat it as a simple 'billiard ball' collision. One only had to know the energy and momentum of the x-ray before and after the collision and the angle through which it was scattered. One also had to assume that the x-ray had hit an electron that was not moving. What a simple picture for the most basic effect in physics!

Richard recommended to Leake that he perform the measurement on beryllium, a light element that did not absorb the x-rays, and promised to send him a crystal of beryllium after he returned to Watertown. On that visit to Cavendish Richard met Heisenberg, who was sitting with Mott in the tearoom. Heisenberg had been interested in the origin of magnetism and Richard's work on the 3d electrons, and was curious to know where Richard was working at the moment. Richard realized that he had moved around quite a bit — the Heisenberg query was no surprise.

Richard had been interested in the old Heisenberg-Goudsmit feud over whether Heisenberg had worked on an atomic bomb during the war, a story that Richard would later dramatize for a BBC presentation. Unfortunately Richard was not then involved in playwriting for he might have had an interesting discussion with the famous Nobel Laureate.

It was now a little over a decade since the promise had been made to Slater to use x-rays to examine electron distributions. During that time the effort had made some inroads although it was clear the physicists still didn't know how to perform the measurement. But they had uncovered some of the stumbling blocks. The neutron results, limited to the magnetic electrons, identified the 'Mickey Mouse' ears to reveal that the electron distributions departed from being spherical. Still, Richard was only 40 years old and there were plenty more mistakes in him. Hopefully, his quest to verify the Schroedinger Equation would still be realized. The nice thing about physics is that when you get it right you sense it. John Leake had started Richard on that path. Success was less than two years away.

WATERTOWN 4: 1964–1972

By the time Richard had returned to Watertown, Chipman and Jennings had completed their measurements on the inert gases neon, argon, krypton, and xenon at 1 atmosphere. At this pressure the atoms are quite far apart on average and could be safely considered to represent isolated atoms. Indeed their electron distributions were measured to be within 1% of the best Hartree-type calculation. It came as no surprise that the Schroedinger equation of 1926 was capable of providing reliable answers to how the electrons orbited about the nucleus of an isolated atom. Any large differences discovered by Chipman and Jennings would have sent the theoreticians reeling and wondering whether anything that came out of Watertown could be trusted.

But having bypassed this hurdle of measuring isolated atoms the problem of electron distributions in solids required attention, both theoretically and experimentally. Humanity lives in a world that is primarily solid and curiosity was aroused over the diverse properties of these materials. For the most part theoreticians only calculated the electron energies of crystals, a quantity not easily measured since the x-rays would have to dislodge the electron from the atom to some final energy that was difficult to calculate. The advantage in measuring electron distributions by x-ray elastic scattering was that the electrons would remain unaltered in the process.

Some progress could be anticipated when Jennings managed to obtain a perfect crystal of copper grown by F.W. Young, Jr. of Oak Ridge. These copper crystals were unbelievably soft and Jennings had to fly to Oak Ridge to personally cradle the package containing the copper in his lap. Jennings was a superb experimentalist who took impressive care with his measurements.

Perfect crystals had an unusual property. It was possible to find directions in the crystal, which were transparent to x-rays. The x-ray waves avoided the electrons that would ordinarily absorb them. This was only possible if the atoms in the crystal were perfectly lined up unlike most materials that contain mistakes like impurities in their atomic arrangement. Unfortunately only a handful of elements such as silicon, germanium, gallium arsenide, and copper could be prepared in perfect crystal form, iron not being one of them.

Jennings went to work and published his results in 1964. His measurements yielded a 3% difference between the best Hartree-type calculation for the isolated atom and the metal. Agreement was also achieved with the BCD measurement on copper powder. What could one deduce at this stage of the Watertown research effort?

1. The measurement techniques were improving and reaching an uncertainty of about 1%.
2. Many of the experimental pitfalls had been identified.
3. The Hartree-type calculations for isolated atoms were in error by 1%, as was expected.
4. The theoreticians were yet to calculate the electron distributions in solids to an error of 1%.
5. None of these results helped the Army make a better tank although some Generals were still waiting.

Jennings' success with a perfect crystal of copper perked up Richard's interest further. He had secured perfect crystals of silicon and germanium and several perfect crystals of GaAs and began to make measurements with John DeMarco. But a phone call in the spring of 1964 from Professor Volker Weiss at Syracuse University sidetracked him,

"Dick, this is Volker Weiss, in Syracuse. We've heard about the terrific work on electron distributions at Watertown. How would you like to hold a conference on the subject this summer?"

"Interesting idea. I'd like to think about it."

"The dates are August 17–21."

"Why is Syracuse interested?"

"We're really not," said Volker, "But the Army has already signed a contract with Syracuse. We have a conference center on Racquette Lake where the Army had planned to hold a conference on the strength of steel. They had to cancel that."

"Where does the money come from?" asked Richard.

"It's all paid for by the Army."

Richard knew it was good fun spending other people's money and he agreed to it. He had already been responsible for a reactor costing more than a million dollars and several van de Graaff accelerators at about a half a million — why not a paltry fifty thousand to hold a party in honor of the electron? Volker Weiss assured him this carte blanche check could be used to bring scientists over from Europe, to provide for their food and drink, and, hopefully, to make everyone happy at the Sagamore Lodge.

The establishment had been built by Cornelius Vanderbilt, the railroad baron. The story had it that he made a wager he could host a Christmas banquet sometime in the 1870's in snow-drifted upstate New York, hundreds of miles from New York City and remote from civilization. Vanderbilt ran a special railroad track from the main line to Racquette Lake and built a comfortable lodge to accommodate about a hundred guests and staff. When the Depression came fifty years later the Sagamore site was threatened with foreclosure for failure to pay taxes so it was donated to Syracuse University, a tax-free institution that agreed to use it for conferences.

The organizers adopted the name THE SAGAMORE CONFERENCE ON CHARGE AND SPIN DENSITY. Charge density alluded to electrons, which were negatively charged and spin density to electrons that possessed magnet moments that could be detected with neutrons. Batterman, Freeman, Nathans, Shull, and Richard served on the organizing committee and proceeded to invite their scientific friends in the field. Sixty people turned up. Except that the random room assignments accidentally placed Freeman and Nathans in the same cabin and necessitated a rapid reassignment, all went smoothly. Scientific sessions were convened in the mornings and evenings with the afternoons free to swim, canoe, drink, and whatever else seemed appropriate to glorify the electron.

A banquet, a string quartet, and a distinguished after-dinner speaker Professor R. Brill wound up a successful conference. The participants were too well 'lubricated' to engage in any scientific disagreements. In 1964 Brill had occupied the directorship of the Fritz-Haber Institute in Berlin, formerly the Kaiser Wilhelm Institute that Heisenberg had headed during the war. The first efforts at building a reactor took place there under Heisenberg's direction but the limited supply of heavy water doomed its chances. Furthermore, Allied bombing of Berlin forced the uranium program to move to a remote site in southern Germany. After the war the name of the Berlin locale was changed to the Fritz-Haber Institute to honor the former Jewish Director who was awarded the Nobel Prize in 1918 for producing synthetic ammonia. He had been forced to leave Germany under the Nazis and died soon afterwards.

A questionnaire sent out after the conference convinced the organizers that they should do it again in three years. Thus, the triennial Sagamore conferences on the misbehavior of the electron would continue unabated to this day although it has shifted its locale around Europe and even to Canada and Japan.

Some fuss was made over Richard with the release of his book *SOLID STATE PHYSICS FOR METALLURGISTS*. He presented a copy to the Commanding Officer, to each of his colleagues, and to his mother who liked the cover. Richard took up residence near John DeMarco in the town of Avon, a 40-minute commute to Watertown. Avon was virtually the smallest town in Massachusetts (2 miles × 2 miles), and had only recently celebrated the 75th year of its divorce from the Town of Stoughton, having had a running battle over water rights and personalities. It needed a name and the natives chose Avon in recognition of the Bard of Avon, the famous playwright and poet who actually was born in Stratford on the river Avon. The visage of Shakespeare was selected for the town seal. The Selectmen recognized the streamlet running through it as the Stratford. Avon upon Stratford had a nice ring to it and should put the town on the map.

Avon's size rated but a small footnote in history. On arrival Richard enquired whether anything of note had ever occurred in this sleepy hamlet but no one knew or cared to remember. Its form of town government entitled every voter to attend the annual town meeting and to make an ass of himself by querying every last penny appropriated for town use. At one meeting the Police Chief asked for $200 for a set of law statutes.

"Will the Chief tell us why he needs an identical set of books as the Selectmen have in the Town Office?" asked one taxpayer.

"Sometimes we arrest someone at two or three AM and we have to know on what charge to book them."

"Why not use the set of statutes in the Town Office?"

"It's locked at night."

"Mr. Moderator, I propose the $200 appropriation request by the Chief of Police be amended to 25¢ for a key to the Selectmen's Office."

The vote in favor of the amendment was unanimous.

The nationally famous Sacco-Vanzetti case struck at the heart of town prejudice. In 1920 an elder in the Baptist Church, working as a guard for the Brink's armored car agency, was killed. Sacco and Vanzetti, two Italian immigrants were apprehended and placed on trial. The townspeople were convinced the authorities had found the right culprits but legal maneuvering and appeals delayed their final execution for almost a decade. The son of the slain Brink's guard was so upset he committed a robbery for revenge. Recently Governor Michael Dukakis pardoned Sacco and Vanzetti claiming they did not get a fair trial. Hence the events of 1920 that touched Avon remained newsworthy for 70 years.

Still, its small size amused Richard. The US Government provided the post office with two mailboxes, one marked AVON ONLY and the other OUT OF TOWN. The regimented town residents dutifully sorted their mail before posting. Richard came upon the post office clerk emptying the two mailboxes and dumping their contents into the same mailbag!

"How come you have two separate mail boxes?" asked Richard.

"Efficiency," replied the clerk as he hastened into the post office.

In the early 1960's high-pressure measurements experienced a resurgence of interest. The field had been dominated by Professor Percy Bridgman of Harvard who had been awarded the Nobel Prize in 1946 for his pioneering work that dated back to the 1920's. Cambridge Massachusetts born and educated, Bridgman spent his entire career at Harvard. During WWII he measured the compressibility of uranium and plutonium, vital to atomic bomb technology for squeezing the material under explosive pressures to achieve a critical mass.

Pressure and temperature were both intrinsic to the field of thermodynamics. It was much easier to control temperatures in the laboratory so that pressure measurements took a back seat. For practically all metals it required a high pressure to effect any changes, say 10,000 atmospheres (an atmosphere is 15 lbs/in^2). Eventually, under laboratory conditions it became possible to reach pressures of several million atmospheres, equivalent to that at the center of the earth, with sample sizes about the size of the head of a pin.

In the 1950's, shortly after General Electric announced the synthesis of artificial diamonds under pressure, Richard spoke to John Holloman who headed the project. Holloman had the distinction of working at Watertown Arsenal during the War, contemporary with Clarence Zener. Holloman assured Richard that in ten years GE would be turning out diamonds as large as one's fist. This turned out to be an overoptimistic prediction to this day. While artificial diamonds were commercially viable for grinding wheels, the subsequent discovery of diamond pipes (veins) in Russia and Australia adequately supplied the market with gem diamonds. The rumor that the South African DeBeers Company was hoarding diamonds to keep up the price was inaccurate although the Russians and Australians did sell their diamonds through DeBeers since they had an established marketing organization.

A small group of experimentalists and theoreticians met in the 1960's to present papers on their high-pressure results. Bridgman joined in as a member of the audience. When Richard suggested that the group try to define a standard for pressure, just as the melting and boiling points of water provided a standard for temperature, Bridgman declared it sounded too much like Government control. A shocked Richard retreated into silence. RHIP (Rank hath its privileges).

Measurements had already begun to examine materials under pressure with x-rays by squeezing the sample between diamonds, the hardest substance known. These experiments could only seek changes in the crystal structure, not in the electron distributions. Richard's pressure interests almost evaporated after Bridgman let it be known that the Government was an undesirable ally — shades of von Hippel.

From 1962 onward John Leake had been laboriously trying to measure the amount of Compton scattering at very low scattering angles. It was known that in the 1930's DuMond at Cal Tech in Pasadena had performed a most exacting measurement on Compton scattering on beryllium by examining the energies of the x-rays that had undergone Compton Scattering. Under appropriate experimental conditions it was this range of energies that could be related to the electron distributions calculated by Freeman and Watson or at least approximately, thought Richard.

The calculations of Hartree and those that followed were aimed at establishing the electron distributions on isolated atoms. However, these same solutions would also provide the momentum distributions. Whatever the electrons did to move from point to point around its nucleus its speed, hence momentum, would be constantly changing. Unfortunately the Schroedinger Equation did not predict the point-to-point electron velocities, only the probability for various values of momenta from zero on up. The reader may be confused about the predictions of quantum mechanics but if one lives with the concept for a while one eventually accepts it. In quantum mechanics both electron positions and electron momenta can only be expressed as probabilities. Einstein never accepted this uncertain aspect of quantum mechanics. He is reputed to have said,

"God does not throw dice

But since Schroedinger's time no one had overcome this limitation to quantum mechanics. It was the only game in town. It could be viewed as a manifestation of Heisenberg's Uncertainty Principle — if you knew the electron's position precisely you were faced with an uncertainty in its momentum. Mind you, the electron probably knows where it is and how fast it's moving but it's not telling. And if you devise an x-ray measurement to locate the position precisely by scattering an x-ray from it you will disturb its momentum and introduce an uncertainty, and vice versa. For years Einstein and Bohr fought over this but Bohr and the Uncertainty Principle always won out.

About this time Paskin nagged Richard to redo the DuMond measurement but Richard was not optimistic that it could be done nor that the answers could be related to the solutions of the Schroedinger Equation. In the Compton Effect the electrons were being dislodged from their positions on the atoms and this would complicate the Hartree-type calculations. This was not a problem in the measurements of the electron distribution since in that case the electrons remained undisturbed.

DeMarco and Weiss began measurements on electron positions in silicon, germanium, and GaAs in early 1964 following a paper published by Richard on diamond. The diamond crystal structure is shared by these crystals. It is non-centrosymmetric, a crystallographer's term meaning it shows a different arrangement when viewed back to front. In the diamond structure there are x-ray scattering peaks that are only present if the electron arrangement is not spherical. This is just what one would expect since each atom has a near neighbor in one direction but not in the opposite direction. It is this electron arrangement in the space between near neighbors that gives diamond, silicon, germanium, and GaAs its special semi-conducting and brittle properties.

To examine the state of the art in theoretical and experimental work on electron distributions circa 1960 Richard compared the values of the x-ray scattering as calculated for crystalline diamond by Kleinman and Phillips and as measured by Göttlicher and Wölfel. The theory could be improved. Richard made an empirical guess as to what the electron distribution might look like in diamond and obtained good agreement with the experiment.

The experimental results of DeMarco and Weiss on silicon, germanium, and GaAs were published in 1964–65 but these could only be compared to Hartree-type isolated atom calculations except for the measured departure from a sphere, a clear indication of the changes expected in a solid. It also was shown how this departure from a sphere could be measured in vanadium metal.

But everything considered it was all rather bland. Until the determination of electron distributions could be measured to reliable accuracy theoreticians were scarcely interested. The entire field needed a shot in the arm. So what does a scientist do when he doesn't know where to find the answers or the path out of the jungle? He writes a book on the subject! His friend Peter Wohlfarth at Imperial College had been appointed editor of a new series of books by North Holland and Richard signed a contract to produce one on electron distributions. This would be an opportunity to tell the scientific community what Richard didn't know, even though authors customarily made it appear that they fully understood their subject.

Between drafting chapters and some further nagging by Paskin, Richard had occasion to visit England for a conference and he arranged to stop at the University of Sheffield where his work on the thermodynamics of iron and the unusual magnetic structure of gamma iron aroused considerable interest. Richard had previously visited Professor John Crangle at Sheffield in 1956, a visit he shall always remember. It was November, a damp, cold month for that city in the Midlands. He was booked into a B&B that failed to provide heating except by insertion of shillings every 15 minutes into a gas heater. Richard had found this heating arrangement totally inadequate. Besides, who wanted to get out of bed to feed the kitty with shillings? It was miserable, shivering between damp sheets and unable to stop his teeth from chattering. But this time his host Professor Norman March booked him into the Grand Hotel and Richard survived without chilling memories.

He still remembers having a few hours to spare for a stroll around the physics building and wandering into an antique shop across the road. There he found an oil painting, not executed in the finest tradition but nonetheless clearly of vintage 1840 and depicting Conway Castle in Wales. Richard identified the edifice he had visited in 1962 and recognized the famous Stephenson tubular railway bridge and the Telford road bridge crossing the river. The trains emerging from the tubular bridge in the painting were circa 1840. It had been a remarkable engineering feat to erect a rail crossing over the River Conway by constructing a long steel tube of square cross section and floating it into place. Such a tubular construction provided considerable strength per weight, the first of its kind. Richard casually asked the price, expecting something like £200 or more to be demanded.

"Three pounds."

"Does that include the new frame?"

The shopkeeper nodded and Richard's purchase has adorned his home ever since. Obviously the proprietor was unaware of the subject matter of the painting. He probably studied physics at the university, thought Richard.

Norman March was a first class theoretician and a lay preacher, somehow managing to keep the two disciplines in proper perspective. The subject of the accurate Compton measurements of DuMond on beryllium was broached. Disturbingly a similar measurement on lithium performed in 1936 appeared to yield a distribution in electron momenta far broader by a factor of two and a half than calculated by March. Now, thought Richard, theoreticians may not achieve the highest accuracies in their calculations of electron distributions in solids and Richard might accept a difference of as much as 20% but not 250%! Here was the enigma that was to put life back into the purposefulness in Richard's life and a renewed chance to fulfill his old promise to Slater. When he returned to Watertown he got cracking, convinced that it would be possible to establish whether the theoretician Norman March or the 1936 experimentalist was correct. Besides, thought Richard, Norman's lay preaching granted him better insight into the work of The Deity.

March was a cautious man. In a previous letter to Richard dated 28 January, 1965 he had written:

"I am very gratified to hear that you have resolved the problem on the Compton profile of lithium for, as you know, it has worried me over a period of ten years. On the other hand a theorist should never say that experiments are wrong though it appears that in fact the 1936 measurements must indeed be. Actually, a day after getting your letter I heard from J.A. Leake to say that his preliminary results also show that the lithium profile obtained by Kappeler is too broad."

In a short internal US Army report Richard confirmed that the 1956 calculations of Donovan and March were in good agreement with Richard's experiment. No doubt the prior experiment by Kappeler had failed badly for some inexplicable reason! Richard contemplated this new development in light of his old measurements on iron and chromium. This time he wasn't taking chances. He wrote to W.H. Taylor and his students John Leake and Malcolm Cooper to describe his own results in some detail. "Why don't you repeat these in Cambridge and we'll publish jointly?" This was agreed by return post! Richard was buying insurance.

As the measurements on both sides of the ocean got underway letters were frequently exchanged to compare results. It took about six months for the dust to settle and for both laboratories to come to full agreement. In a letter from John Leake dated June 21, 1965 he confirmed that the Cambridge and Watertown results closely agreed and he opened discussion about the mechanism for a joint publication.

And so it came to pass that in the Philosophical Magazine of October 1965 a joint publication by Cooper and Leake from Cambridge and Richard Weiss from Watertown revealed that they both measured the Compton profile of lithium and had obtained identical results, both in agreement with the 1956 calculations of Donovan and March. Success was sweet but Richard did not fully realize the rewards awaiting him on this new adventure into electron momentum distribution.

By now Richard had finished the manuscript for his book on electron distributions. It would still be a few years before the extent to which the electron momentum distributions would prove superior to electron position distributions. Hence the book *X-RAY DETERMNATION OF ELECTRON DISTRIBUTIONS* appeared in 1966, a few years before it could include details of the momentum distributions.

Two books in three years, Richard's publishers sent him copies of the reviews including one in Russian, which was a translation of one in English, and one in English, which translated the Russian review back again. The reviews of Richard's earlier book on solid-state physics were generally favorable and sympathetic to his new pedagogical approach, but there was only one point on which the reviewers agreed on the latter book — it was too expensive.

Richard had rarely read a review that convinced him to go out and buy a book unless it was clearly something he would have purchased anyway. Book reviewers were underpaid, receiving only a complimentary copy of the text. Trade book reviews published in newspapers, like movie and stage reviews, were read for their literary style, their satire, their barbs at the author, and anything else that demonstrated the reviewer to be a vastly superior individual to the author. To a lesser extent this was true for scientific books but it's still done to excess. It is rare to find a book review that does not point out some trivial error or mis-statement or what have you.

A fresh face appeared on the scene. Walter Phillips, quiet and unassuming, had taken a job with NASA in Cambridge before their new building had been renovated. He was given carte blanche to work elsewhere until his office was ready and he approached Richard to offer his services. He was familiar with the new work on Compton Profiles and agreed to a joint effort. Richard pointed out that they would have to start afresh and that it would likely take a few years to solve the experimental and theoretical problems. The x-ray intensities would be weak and it would require long automated counting times. Besides, there were bound to be unexpected difficulties and there was no assurance these could be surmounted. Nonetheless, Phillips accepted the prognosis and the hard work began.

As this experimental effort proceeded Richard began to consider the organization of the 1967 Sagamore Conference. The organizing committee was selected to include both geographical and scientific considerations; Barrie Dawson from Melbourne, George Hall from Nottingham, J. Ibers from Northwestern, R. Watson from Brookhaven, and Cliff Shull from MIT as chairman. Richard was relegated the position of secretary which meant he had to take care of all correspondence, raise the money, solicit speakers, and deal with the Syracuse University administration. While Walter Phillips and he were attempting to measure the electron's momentum in the adjoining laboratory Richard's own momentum in and out of his office rarely dropped to zero. Phillips arrival turned out to be a godsend.

In 1966 Richard received a letter from Princeton to advise him that the Rockefeller Public Service Awards were entering their 15th year and they were soliciting written comments from the 79 awardees to present to John D. III. Richard sent the following reply in October:

Mr. John D. Rockefeller III
Rockefeller Public Service Awards
Princeton, New Jersey

Dear Mr. Rockefeller,

It makes me most happy on this special occasion to add this note in recognition of your foresight and contribution to our Government. The Rockefeller Public Service Awards, established in the highest tradition of philanthropy, have a very special meaning to those in Government service much as the Nobel Prizes have a significance to all.

My own award, enabling me to spend a year at the University of Cambridge, was received ten years ago and rather than being 'forgotten' as an event of the past, its value has steadily increased. Over and above the benefits derived from study abroad, the knowledge that I am one of an extremely select group who have served their Government with distinction has developed in me a very special attitude towards Government service. I am devoted to the cause of improving Government service to rival the finest parallel service in academic and university establishments; and I can point with pride to this laboratory's worldwide recognition in physics.

As the years pass, the prestige attached to these awards will continue to grow. As an amateur historian I venture to guess that ultimately the Rockefeller name will not be primarily associated with oil or politics but with these awards.

Sincerely yours,

Richard J. Weiss

It's difficult for Richard to believe that he could have written such insincere drivel except in an effort to use flowery language where a simple 'thank you' would have sufficed. Richard attended a presentation dinner at the Shoreham Hotel in Washington on April 12, 1967 where the letters from recipients were presented to John D. III.

The 1967 Sagamore II conference was held September 5–8 at the same site as Sagamore I, a week before the International Conference on Magnetism held in Cambridge, MA. It was decided to add electron momentum density to the topics under discussion although it was too early to add much solid information on this subject.

A good time was had by all particularly since the Sagamore conference policy at the time did not require any of the attendees to write up their talks. The booze flowed, the lake splashed to the sound of swimmers and canoes, and the otherwise sedate Cliff Shull surprised us all by delivering a riotous after-dinner talk. Richard was further pleased to have met the famous Paul Ewald, one of the pioneers in crystallography. The tall, imposing, and kindly gentleman was born in Berlin in 1888, seven years before the discovery of x-rays. He could relate many a tale of the golden days of crystallography and his early theoretical work left the field with the designation THE EWALD SPHERE to indicate which x-ray reflections came into play under specific experimental conditions. As another attendee humorously whispered to Richard about this memorable figure,

"The most perfect geometrical shape, the sphere, was named after Ewald. What greater honor can befall anyone!"

Richard ran into Paul and Mrs. Ewald at crowded Heathrow Airport a few years later, after he had reached his 80th birthday. Richard was taken aback to find him in a wheelchair being trundled along by Mrs. Ewald.

"Hello, Paul. Anything wrong?"

"I'm alright, Richard. It turns out that the wheelchair is free and it is an easy way to get around the airport when it's difficult to find a seat."

Paul left Germany of his own volition during the Nazi period. His daughter married the well-known nuclear scientist Hans Bethe. He studied with the great Arnold Sommerfield in Munich and was a contemporary of von Laue. Paul helped launch   the   standard   journal   on   crystallography   *ACTA CRYSTALLOGRAPHICA*.

Erwin Bertaut of France came forward at the end of the conference and offered France as host to the 1970 Sagamore III, to be funded by their national research organization CNRS. It was eagerly accepted by the organizers.

Returning to Watertown the Phillips-Weiss duo became entrenched in the Compton profile measurement. Their first results on lithium and beryllium, now much improved over the earlier work on lithium, placed Richard into a puzzling dilemma over interpretation. He paced the corridor for miles searching for a way out.

In measuring the electron distributions the x-ray bounces from the sample in all directions without change in energy. But in the Compton effect the billiard ball collision with a single electron sends the electron off with increased energy and the x-ray with decreased energy (the total energy is conserved). If the electron was initially at rest (zero energy) the x-ray would bounce off with an energy loss that only depended on the angle of the collision. However, if the electron had some momentum before the collision the scattered x-ray displayed a range of energies, which, hopefully, one could relate to its initial momentum and to the solutions of the Schroedinger equation. Richard tried guessing what happened to the electron after the collision but discovered that the collision between the x-ray and electron was so rapid (speed of light) that it was unnecessary to consider all these possibilities. It was as if the electron did not care a whit what happened to it until after the x-ray had disappeared into space. This became known as the Impulse Approximation and provided the cornerstone for analyzing the Compton effect. It was a blessing of untold significance for it made the measurement easy to interpret and readily related to the mathematical solutions to the Schroedinger equation. Where Richard's years of difficulty in measuring the electron distributions relative to the nucleus uncovered mounds of difficulties the electron momentum distributions proved to be most revealing.

What nature hid in the electron positions it shouted out for the electron momenta.

And what Jesse DuMond had 'roughed out' in the early 30's the improved x-ray equipment of the 60's spelled out in minute detail. At age 77 DuMond wrote Richard:

"My first student H.A. Kirkpatrick and I were the first to study the structure of the Compton line and the electron linear momentum distributions. This work was my "first love," as we used to call it. I shall never forget the "kick" I got when, one Sunday, I was so impatient to know the results of one of our long exposures with the 50-crystal spectrometer that I couldn't wait until Monday. I developed the film and turned on the ruby light and, sure enough, just as my theoretical reasoning predicted the width of the Compton line was much narrower than at 90 degrees. I KNEW SOMETHING FOR CERTAIN AND, FOR THE FIRST TIME, THAT NO ONE ELSE IN THE WHOLE WORLD WAS SURE OF OR EVEN BELIEVED. The elation of such a moment is indescribable! Thereafter I had to be prepared for a lot of disillusionments about people, physicists, and other human beings, who refuse to believe one's interpretations, who are too lazy to read one's evidence, but the 'kick' of that one experience, all alone on a Sunday, carries one through a lot of these 'realities of life'."

No doubt DuMond had to run through a line of disbelievers! Richard's discovery of the Impulse Approximation unearthed a similar response. Nature's fundamental secrets were revealed to but a few.

In Richard's reply to DuMond he stated:

"For the first time we now have a sensitive test of solid state wave functions and this puts the experimentalist a "giant" step ahead of the theorist."

All this almost coincided with man's "giant" step onto the moon.

The Physical Review of July 1968 carried the details of Phillips' and Weiss' effort in the reincarnation of the DuMond work some 35 years before. It provided one of the most sensitive tests of the Schroedinger equation and soon a myriad of experimentalists and theoreticians were enjoying this excursion into quantum mechanics.

Not only that, but Compton Profiles clearly demonstrated that the theoretical model of a metal was consistent with the concept of a Fermi momentum, i.e. the current-carrying electrons had a range of energies up to a maximum called the Fermi energy. Up until the time Walter and Richard measured it, it was a vague theoretical notion. But to actually see it for the first time, as indeed DuMond had related in his letter upon discovering the momentum of electrons in beryllium, was a delightful surprise. Too bad Fermi had passed away in 1954 he might have been pleased to hear about it.

One person delighted to learn about the successes in Compton Profile measurements was Professor C.A. Coulson, F.R.S. at University of Oxford. He had performed some calculations on the momentum densities of molecules before WWII. Richard wrote to tell him that he had measured some of these molecules. Coulson wrote on 16 April 1969:

Dear Dr. Weiss,

"Thank you very much indeed for sending me the preprint of your paper on the Compton Profiles of benzene and some other hydrocarbon molecules. It made me feel very happy to realize that now again people are becoming interested in these profiles and the information which they provide about momentum distributions. The work by Duncanson and myself to which you refer was done nearly thirty years ago and still we are not certain about what the momentum distribution really is. So you see I read your work with a great deal of interest."

Yours sincerely,
s/ C.A. Coulson

Sagamore III at Aussois, near Grenoble, added momentum density to its title and it has since been designated as the conference on charge, spin, and momentum density. Bertaut organized the conference in the best tradition of ethanol lubrication. French wine gave one a real <u>charge</u>, sent one's head into a <u>spin</u>, and provided ample <u>momentum</u> to get to the bar before the wine ran dry. Except for the beef Bourguignon which made a few Americans ill from overindulgence the conference was a great success. Richard remembers Bertaut reminiscing in his French-English about the summers at Sagamore I and II and the beautiful Adirondack surroundings,

"But you had to be careful when you walked in the woods — you might run into a beer (bear)."

Richard added, "On a hot day you might want to run into a beer."

144

WATERTOWN 5: 1970–1979

As useful as the Compton Effect had been in illuminating the intimate details of electron behavior it had to undergo a Caesarian birth in 1920, having to be torn from established ideas and prejudices. Professor Roger Stuewer, an outstanding academician in the history of physics, and Professor Malcolm Cooper, one of the pioneers in the subject's rebirth, pieced together this fascinating story for a book on Compton Scattering published after Sagamore V.

In their introduction they say:

"Both in the period leading up to the discovery and explanation of the Compton effect in late 1922, and in the period which followed, when its potential as a method of investigating the behavior of electrons in solids was realized, there were many controversies, numerous inconsistencies, and a whole shoal of scientific red herrings! There were early spells of intense activity by a few individuals — and later, an unbelievable interlude of a quarter century of complete inactivity. Those who believe that scientific discovery and innovation ought to progress smoothly and logically will find scant comfort in the story unfolded below."

The dilemma that faced the physics community at the turn of the century centered around the dichotomous wave and corpuscular theories of x-rays — were they the one or the other? Only after accepting the dual nature of x-rays, i.e. they were both, was man's limited power to comprehend able to overcome the difficulty. Once Compton recognized this and accepted the momentum of an x-ray as given by its energy divided by the velocity of light, the simple picture of a billiard ball collision between x-ray and electron fell into place. The next giant step was taken around 1930 by DuMond and Kirkpatrick in measuring the energy-broadened Compton line to show that the electron had momentum before the collision. While this presaged great things for the field, for the next 35 years the measurements took a back seat because the DuMond experiment was so terribly difficult and nuclear physics emerged to set the fashion for the scientists. After this hiatus, broken by the Cooper, Leake, Weiss publication on lithium, Richard recognized the validity of the Impulse Approximation and this enabled the subject to acquire a firm experimental and theoretical basis. In a few years it soared to new heights.

Following the 1970 Sagamore III Conference in France Richard polled the attendees and asked their preference for the 1973 site. Both the Finns in Helsinki and the Russians in Minsk were anxious to host the conference and it befell Richard to choose. The Finns were the emotional favorites but the International Conference on Magnetism would be held in Moscow in 1973 and there were several individuals who wished to attend both. Richard opted for Russia with the understanding that it would move to Helsinki in 1976, America's bicentennial year. The Finns were not happy with Richard's decision and several probably never forgave him. The Finns hatred of the Russians rivaled the American Patriots' animosity toward the Redcoats in 1776.

In the early 70's Richard kept up with news from England by subscribing to the Sunday Observer and reading the New Scientist, a popular London-based magazine that detailed current scientific news for the general public. The editors invited letters from the readership in response to their articles. In the satirical tradition of America's first physicist Benjamin Franklin, Richard sent the following to the editor:

Sir,

Peter Browne's article (The Llandudno Pentagon) supplements the well documented disappearances into the Bermuda triangle and should encourage scientists to seek within the British Empire areas of other symmetries, i.e. the square, the hexagon, etc. However, I urge those so inclined to look particularly for disappearing areas of lower symmetry, i.e. the line, the point, and most important of all — that of zero symmetry the 0 (pronounced like three Welsh double ll-ll-lls). Because it is into the land of the 0 that the falling pound is being inextricably drawn. By plotting the value of the pound as a function of time (from 4.80 in the 1920's to the present 1.7) one can extrapolate its disappearance into 0 at about the year 2000. This gives us some 25 years to solve this mystery and save the pound.

Incidentally, the Editor is no doubt aware of a similar pentagon in the US into which huge sums of money have disappeared.

R.J. Weiss
Watertown, Mass.

The Blanchard Trust came to Richard's attention during this period. Henry Lawton Blanchard left a goodly sum in trust requesting the trustees to use the money to interest people in the lives of their forebears. Nothing had been accomplished since the Trust's inception in 1947 and Richard suggested that a miniature village, like Madurodam in Holland, be constructed on an 8 to 1 scale to model Avon from its first settlement to the present. Madurodam had fascinated Richard but the trustees were not impressed and the idea fell by the wayside. Still, after 25 years the trustees must have felt some obligation to implement Blanchard's will. Richard decided to wait a few years and nag again.

In 1972 negotiations about Sagamore IV began with Academician N. Sirota in Minsk. When Sirota had attended Sagamore II he had been invited to speak about the work of his Institute and he began his talk in a thick, hesitating, but understandable accent:

"You must excuse my English because it is so beautiful!"

Richard raised some administrative points in his correspondence with the 'beautiful Sirota' but quickly discovered that the Bjellorussian Academician was accustomed to being boss. Richard decided to let Sirota carry the ball alone — on his head rested the success of Sagamore IV and the reputation of the

Russians. Why worry about something behind the iron curtain that Richard could do nothing about?

Actually it was a bit bizarre at that period of the cold war for an employee of the US Army to be arranging a conference behind the iron curtain. Richard did not advise the security officer at Watertown of this development — it would have been impossible to justify his actions! Richard recalled a prior visit to Professor Rolf Hosemann at the Fritz Haber Institute in Berlin when he asked about visiting East Berlin. Since no German resident of West Berlin was permitted into East Berlin at the time, Hosemann assigned one of his Spanish students with a Renault Quatre Chevaux, an extremely small auto, to take Richard through Checkpoint Charlie. Riddled with uncertainty as to whether he was permitted to visit East Berlin Richard enquired of the American sergeant on duty as to the rules.

The sergeant rang up headquarters which asked for Richard's Civil Service rank and duties. When he informed them he was a physicist GS-15, the highest rank for non-presidential appointments, Richard was ordered to stay out of East Berlin.

"They might grab you and squeeze your balls for your secrets," said the sergeant.

No use arguing, thought Richard, with such lame excuses like Richard knew no secrets, nor that nothing on his passport indicated his status with the US Army. He and the Spanish student returned to Hosemann's laboratory where everyone had a good laugh. If the East Germans had refused entry they would have understood, but for the Americans themselves to have excluded Richard was much too funny not to make the rounds of after-dinner gossip.

On a previous visit to East Berlin the Spanish student told Richard he had been stopped on his return and his small auto was thoroughly searched, including poking a stick into the gas tank.

"What are you looking for?" asked the student.

"I'm looking for the tiger."

East German guards were notoriously devoid of a sense of humor. Here was one who broke the mold.

Hosemann escorted Richard to several points along the Berlin wall where platforms had been erected to view the hideousness of the no man's land that kept their citizens from escaping. At one point the wall separated one half of a street from the other half. Peering down the road Richard saw a youth of about 10 staring at him. By the fickle finger of fate this poor lad grew up imprisoned, possibly robbed of his former playmates who lived at the other end of the street. It was too much for Richard to bear — he turned away from this dismal example of man's inhumanity to man.

In order to get to Minsk it was necessary to fly via Aeroflot from East Berlin. Richard had arranged to meet Hosemann in West Berlin so that the two could travel together from Schönefield Airport in East Berlin. At that time visas were required for visits to Russia and Richard was informed by several people who had toured Russia that Intourist, their official tour agency, had to make the arrangements and this required prepayment of hotel bills, travel fees, etc. before the visa was granted. Richard ignored this advice and merely sent his passport to the Soviet Embassy with a copy of Sirota's invitation. He received the visa by return mail proving that Academicians had clout in the USSR.

Richard left Boston on August 11, 1973 and flew to London on the overnight TWA 754. After a three hour stopover at Heathrow he boarded a Pan AM flight to Berlin arriving at 2 PM. Hosemann picked him up and they drove to his home in Dahlem where the overheated Richard was led into the garden, had his clothes removed, and placed under an outdoor shower, all in ten minutes. Refreshed, Richard dressed and was introduced to Frau Hosemann and one of Rolf's sons.

"What do you do?" asked Richard.

"I'm an art student."

"You decided against science?"

"One scientist in the family is enough, as long as he's working!"

Richard got along well with Rolf Hosemann, a former wartime officer in the German Navy. They both smoked cigars, an additional bond of friendship although Rolf had the disturbing habit of using snuff (schnupf auf Deutsch). In addition, he would virtually force his guests to indulge by bringing out his schnupf collection. This consisted of a large wooden humidor with bottles numbered and identified from 1 to 12 as well as Rolf's own blends like 1&4, 2&7, or 3,4&9. No man entered the Hosemann home without removing his shoes and donning slippers furnished by the host, and then schnupfen.

Richard was reminded of the 19th century sea captain who had a medical kit containing bottles labeled from 1 to 20 with a guide listing symptoms for which each bottle was a remedy. One crewman required #8 which the Captain discovered to be empty. He substituted half a dose of #3 plus half of #5. The man was still cured!

The morning after Richard's arrival Rolf ordered a taxi and the two passed through a border search at a rarely used gate into East Berlin and then moved onto Schönefield air terminal. One story making the rounds concerned an American scientist who entered Russia through Moscow and was advised that the copy of Life magazine he carried was considered contraband. He was told to remove it upon his departure from Russia. Two weeks later he left via Leningrad and was asked to produce his copy of Life magazine! Russia and East Germany maintained a very large police force — one gendarme watched two citizens while the two citizens watched each other.

Richard was struck by the primitive East German facilities (compared to an American airport). In addition, the plane appeared ancient and the flight to Minsk was unnerving as the pilot came in at treetops. They were met at the airport by Academician Sirota and several of his staff to be driven into Minsk and the modern Hotel Jubileinaya, the site for accommodating the guests and the conference.

Minsk, a thousand year old city of about one million, was the capital of Byelorussia, the most western of the republics, where tractors and trucks were a major industry. Richard was struck by the absence of billboards in this non-market economy although huge pictures of Lenin were often displayed. To Richard it would be like posting a picture of George Washington on every billboard in America. The city had a student population of 100,000 reminding one of Boston where students comprise a significant fraction of the populace. And like most Russian cities open spaces were everywhere, gardens, squares, wide boulevards, etc.

There was the usual formal opening session at the conference at which at least one party apparatchik spoke to extol the virtues of something or other. Following this introductory session some entertainment was arranged. Several performers dressed in native costumes appeared with balalaikas and guitars and Richard was prepared to enjoy some authentic folk music. When they broke into loud rock and roll it was too much. No doubt Sirota wanted to please the visitors by playing what he thought they would like to hear.

At the formal scientific sessions translators were employed. Seated in a separate booth high above the audience they did remarkably well in translating complicated scientific phrases. When Richard delivered his talk he began by congratulating the interpreters. They suddenly went silent at this unexpected turn of events. Nobody had ever taken note of them and they were clearly uncomfortable!

Sirota did follow the Sagamore practice of morning and evening sessions with afternoons for recreation. It was at this point that Richard objected to an afternoon visit to view a huge war memorial called Glory Mound, a hundred foot high obelisk-like shaft of granite encircled with a 50 foot diameter granite ring. Richard decided to remain behind and have a chat with the interpreters. Sirota sent a messenger to find Richard,

"Dr. Weiss, the bus is waiting for you to visit the war memorial."

"Please inform Professor Sirota that Richard already saw the real thing."

With Germans and Japanese present Richard considered such reminders of the past as inappropriate to a scientific conference. No doubt Sirota was merely carrying out the party line but as spiritual leader of the Sagamore conferences Richard acted as his conscience dictated. Whether this angered Sirota Richard never found out although they continued on amicable terms.

Several of the conferees spent Sunday morning looking for a church. The taxi drivers were uncooperative but at last someone managed to locate one on the outskirts of town. What a nostalgic sight to see the sanctuary crowded with parishioners. Old women wearing their babushkas, a few old men that survived the war, and even a few young couples that bucked the party line, all holding candles and fervently joining in the service. Arthur Freeman went even further by looking for a synagogue but did not report on his success.

On another afternoon Richard asked the interpreter assigned to him whether there were any philatelic shops. No sooner was the request made than a car arrived and Richard and interpreter were driven to a small open square with a hundred people milling about. All philatelic swapping, etc. was done in this open market. Richard approached one of the merchants who opened his stock book and showed Richard a few pages of stamps. The Russians issue hundreds of new stamps every year and Richard soon realized that the cost of cancelled stamps was small.

"How much for the whole stock book?"

Having posed this question Richard was quickly surrounded by at least fifty dealers taking note of this American engaged in a wholesale trade. The transaction was completed and Richard exhausted his supply of rubles for the day. The dealer added in hushed tones,

"Here is my address. I would like to exchange American stamps with you but you must send them in a plain envelope — trading is strictly illegal."

Richard never bothered to follow up any 'black market' philately.

The interpreter, a pretty girl of about 30 called Lucy Ludmilla, was a student in Sirota's Institute. Her English was passable.

"What do you read to keep up your English?"

"Mostly Charles Dickens and Mark Twain," she replied.

"Why don't you let me treat you to a subscription to a good English newspaper like the Observer?"

She was enthused with the idea and provided Richard with her mailing address. The next day she appeared with a pained look on her face. She took Richard aside and pleaded with him not to send the newspaper. She was young and overnight had learned the facts of Russian life.

On an afternoon stroll Hosemann and Richard found a sweater lying on a park bench. They picked it up and looked about for the owner but no one was there to claim it. A few minutes late a young woman with despair registered on her face came dashing down the path and Hosemann held up the sweater for her to see. She reacted like someone let out of prison and accepted the garment, kissed both Hosemann and Richard, and happily departed. The sweater represented several months' salary.

The scientific content of the work at Sirota's Institute was disappointing in its limited incursion into new ideas. But none of the westerners felt the visit a waste of their time. There were too many other diversions to amuse them.

Freeman approached Sirota and needled him since he had not arranged for the conferees to visit his Institute, emphasizing that such is always done in America and England. Sirota was in an awkward position since such visits had to be arranged through the authorities in Moscow but he threw caution to the winds and immediately ordered a few cars for transportation. An hour later those of us who could be rounded up were ushered into the main laboratory of Sirota's Institute where we were presented with vodka and chocolates or caviar. Sirota and his staff stood by awaiting developments.

The usual welcoming speeches and toasts followed. Richard spoke on behalf of the Sagamore committee and this was translated by an interpreter whose mien could be described as 'stonefaced.' After two humorous stories Richard noted that he did not crack a smile.

"Are you the interpreter?" asked Richard.

The poker face did not alter as the man nodded.

"Well then, when I laugh, you laugh!"

The tour of the lab revealed a modestly equipped establishment engaged in unexciting research. As the termination of the visit someone suggested we call the conference Sagaminsk and the subsequent one scheduled for Helsinki Sagafinnsk. Actually, the Finns were delighted to learn that the smooth-running elevators at the hotel were made in Finland.

Richard discussed Watergate with the Russian conferees but they couldn't believe that it represented anything more than political maneuvering. Nixon was considered a hero by the establishment and there was no altering the party line.

Richard complained to Sirota that the Russians had translated his book without his approval and with nary any royalty. The same afternoon one of Sirota's assistants approached Richard and handed him an envelope — Sirota had dug into his own Institute account and gave Richard $150 in rubles.

"Please spend this in Russia," was the instruction.

Actually, rubles could not be taken out of the country and Richard had no alternative. He therefore invited a dozen of his friends to a scientific session at the hotel. The group met that evening in a private dining room and everyone ordered vodka. When 12 uncorked bottles of vodka were served Richard's rubles disappeared. SOLID STATE PHYSICS FOR METALLURGISTS turned into LIQUID STATE PHYSICS FOR RUSSIAN PARTIES.

The final banquet and its 150 attendees fulfilled every image Richard had been fed over the years concerning Russian behavior on these occasions. Toasts were proposed every 15 minutes with an amber colored vodka that tasted like brandy. It was very smooth and the entire bottle that Richard drank left him with nary a hangover. Richard tried a few stories out at the banquet but Sirota countered by telling Richard to take his stories back to Texas!

As Richard had predicted the conference was a success. As a parting gesture Sirota met the group at the airport to shower each with a gift and to have a final drink. He demonstrated the pecking order enjoyed by Academicians when the plane's pilot entered the departure lounge and approached Sirota,

"Professor Sirota, the plane is scheduled to leave."

"You wait. It will leave after I've said goodbye to everyone."

The pilot nodded and walked to the corner of the lounge to await Sirota's pleasure. Sirota had set a new standard for the Sagamore conferences. But the Finns still had three years to make preparations to top the Russians.

And so it was on to Moscow to attend the International Conference on Magnetism. Richard was booked into the Hotel Rossia, probably the largest such edifice in the world and a showpiece for foreigners. It was referred to as the Comrade Hilton and sported a large rooftop restaurant with a sentry that refused Richard and other Americans admittance with the phrase 'Komplet' signifying it was all booked. Looking through the door there appeared to be an ample number of empty tables. When Richard's colleague Peter Wohlfarth emerged to head for the loo, Richard asked how he gained access.

"Easy. You have to speak German to the doorman."

Richard couldn't understand the reason for such policy but who could figure out the Russians?

Bathtubs and sinks were devoid of stoppers — it was an item that could not be kept from pilferage. The Russian economy did not provide for an ample supply of these items in the shops. Soap and toilet paper were doled out to each patron by the concierge on each floor.

One bit of trivia that Richard shall not forget during this summer of 1973 was the emergence of Solzhenitsyn as a world figure in Russian literature. By a quirk of nature he happened to closely resemble Richard and this was pointed out almost daily in his interaction with people on the streets of Boston. But not once during his sojourn in Russia did anyone mention this. No doubt, as persona non grata, Solzhenitsyn's picture was kept out of the Russian newspapers. One Muscovite did stop Richard in the street to ask directions in Russian but Richard made it clear her was an Amerikanski.

One of the amusing aspects of Russian life was the presence of middle-aged women with straw brooms who kept the streets clean. These sweepers were in a position of authority and Richard spied one citizen being given 'what for' when he dropped a piece of paper on the street. The task for these women was eased both by the meticulous care taken by the citizenry as well as the absence of domestic pets. Not a single dog or cat was observed during Richard's perambulation of the streets of Moscow.

Someone mentioned an interesting excursion to a working monastery 40 miles outside of Moscow at Zagorsk. The Intourist office ran a tourist bus for a daylong visit and Richard joined some 20 others for this adventure. As the brochure stated:

"Zagorsk is a small town outside Moscow, not to be found on every map. Nevertheless it is the pride of every Russian and a veritable tourist Mecca."

The tour guide was a rather pleasant young woman whose command of English was impressive, particularly her ability to ward off embarrassing queries about life in Russia. She warned the passengers about taking photographs from the bus since it passed several military installations.

"May we look out the bus window?" enquired Richard.

"Yes, but forget what you see."

Such evidence of candor and humor was rare.

The 15th century walled town was a gem, indeed. It was a showplace maintained in pristine condition with gold-leafed onion domes reaching high in the air and its cathedrals displaying a myriad of icons arranged in several tiers. Nary a spot on the interior walls was not covered with religious paintings. Black-robed priests could be occasionally seen moving between buildings. Zagorsk's beauty was even recognized by Lenin who proclaimed on April 20, 1920 that it should be preserved.

How did this support of a monastery square with the official atheistic policy of the state? Quite simply it was called a museum. The guide emphasized that the Communists could not deny past history and it was preserved for all to see. Richard thought it bizarre when one considered the attractions that lured the Muscovite populace. First was Lenin's Tomb with long orderly lines wending around Red Square. But second were the Tsarist artifacts within the Kremlin such as the Royal carriages and jewels. In fact Richard was informed several years later that church attendance in Russia had already reached over 20%, higher than any country in Western Europe.

During the course of the conference Richard was approached by someone who handed him a note containing an invitation to visit Academician Ageev at his Metallurgical Research Institute. Richard hadn't a clue what it was all about but he was game to try anything. He was picked up and chauffeured to a modest-sized building and ushered into a large office where he was introduced to Ageev, offered a drink, and, through an interpreter, began a discussion about gamma iron, Ageev's scientific interest. This three-way conversation lasted about an hour and was followed by a rapid tour of the laboratory. When Richard returned to Watertown he was informed by Dr. Larry Kaufman that Ageev spoke very good English. Why the sham? Richard could only deduce that national pride and the party line forced Ageev to engage in this charade. Russians could not be seen being too chummy with westerners.

Academician Peter Kapitza was a legend in the world of physics, particularly low temperature and high magnetic field research. Richard was anxious to get a glimpse of him at the opening session of the Magnetic Conference. During Richard's stay at the Cavendish Labs in 1956–57 Kapitza's name was often mentioned. He had emigrated to England during the Lenin uprising in Petrograd and took up a position under Rutherford in Cambridge. He would produce very high magnetic fields for a few milliseconds by discharging a bank of condensers through a magnetic coil. In 1929 he became the first foreigner in 200 years to be elected to the Royal Society.

Kapitza made it a habit to visit Russia every summer to see his mother but in 1934 Stalin decided to keep him. The English crated his experimental equipment and sent it to him after the Russians paid all expenses. During WWII he refused to work on the atomic bomb and was placed under arrest but following Stalin's death he was appointed director of a prestigious research institute.

He was released from 'bondage' and visited the USA in the early 80's after wining the Nobel Prize. Richard recalls the story he told at Harvard to demonstrate the advantage of the Communistic system. A fine technician at Cavendish had left his job to run several butcher shops left him when his uncle died.

"This would never have happened in Russia!" he assured the audience. "Society lost an excellent technician."

On his way to the USA Kapitza stopped to visit his old haunts at the Cavendish Lab. At dinner in his old college he realized that he was missing the gown that everyone donned for their meals. He asked the porter to find him a gown and ten minutes later the porter returned with Kapitza's own gown that was still hanging on the cloakroom hook where he had left it almost fifty years before!

1973 found the Russian Jews in a mass effort to emigrate to Israel. As one of these scientists, Dr. Azbel, explained to Richard,

"It has nothing to do with religious preference. I don't go to any formal church. It's just that my heart is not in Russia even though I grew up here."

On the other hand the Russians claimed that they had educated the man at public expense and they could not see him going over to the enemy camp. Actually, the Russians were concerned that a mass exodus of scientists might occur from their laboratories. It became official policy that as soon as a Jew asked for an exit visa he was discharged from his job and banned from further employment. To keep up with their science these 'refuseniks' would hold meetings in each other's apartment. While in Moscow Richard was invited to one of these sessions but he declined since he held some official status with the Sagamore Conferences and didn't wish to antagonize anyone. It took a few years for world pressure to induce the Russians to release these Jews. Many, including Azbel, ended up in the USA.

On the BEA return flight to London Richard breathed a sigh of relief. It was a typical response for westerners when they left Russia. The flight attendant told him the story of two BEA pilots who were warned not to say too much in their Russian hotel rooms since they were probably bugged. These pilots decided to look for the bug and pulled back the carpet to find a metal plate screwed to the floor. They managed to unscrew all the bolts only to hear the chandelier in the room below crash to the floor.

Richard returned to Watertown to continue his work on Compton profiles. By this time there were at least ten groups in the world engaged in these measurements. Both the theoreticians and experimentalists were actively attuned to electron momenta providing a healthy interaction as one learned from the other. It was almost 50 years since the Schroedinger equation came on the scene and it was now being tested to the limit. It had been a brilliant discovery in 1926 but after being checked for electron energies, position, or momenta it was milked dry.

The Boston Globe ran a Massachusetts Science Fair every year at MIT as a final runoff for students who had taken top prizes in regional fairs. Richard served on the committee to oversee the operation and to act as a judge at the fair. This was a delightful duty since these high school students were uninhibited by excessive book learning and the projects they worked on were often more imaginative than dreamed up by professional scientists. There were the perennial experiments to demonstrate that talking to plants increased their growth rate or that tobacco smoke could kill plants, and whatever else captured their fancy.

Richard still remembers the ninth grader who took an interest in spiders. He carefully weighed the fruit flies he fed them and demonstrated that the entire mass ended up as spider silk, i.e. the spider left no waste material on the walls as flies do. Not only was its waste material converted to silk but it could climb the thread and ingest the silk to re-store it in its sac, to be reused to spin its web. In short, the spider invented bungee jumping.

Many interesting questions arose about the material properties of spider silk. The waste material was stored as a liquid but when spun as silk it quickly hardened. If one could weave spider silk as a material it would demonstrate a direct commercial conversion of fruit flies to cloth. Unfortunately, spiders were loners and could not be colonized like silkworms.

Richard read of one biologist who had made an effort to upset the spider's circadian rhythm in hopes it would spin during the day rather than at night. He fed the creature the drug mescalin but failed to alter its web-producing hours. Rather, the spider spun webs of crazy design.

Like many 'living' materials which nature created for strength, wood, bone, hair, silk, fingernails, etc., the molecular structure of spider silk was aligned along the stress-bearing direction. And the longer the molecular chains the higher the viscosity. Undoubtedly as the spider spins its web the long chain molecules are 'polymerized', much as the dentist accomplishes when he employs UV curing for polymeric fillings. The spider seems to be able to reverse the process when it ingests its own silk, converting the high viscosity silk to a low viscosity liquid to be restored in its sac.

It is interesting that spider silk is stronger than steel. If one could synthesize spider silk and store it in a vat, one might have an inexpensive commercially interesting product. However industrial polymeric fibers like kevlar can be fabricated that are as strong as spider silk although more expensive.

One of the concerns of the Science Fair Committee was the sparsity of women in the field. Richard discovered a female engineering Professor at MIT and arranged to see her. He was impressed with her and proposed that she speak to the science fair contestants about the attractions of a science career especially for women. She agreed to give a formal talk to the assembled students and their teachers. No doubt this had some effect on the group. The Professor, Sheila Widnall, had later become dean at MIT and eventually Secretary of the Air Force under Clinton.

In 1974 Richard made another effort to induce the Blanchard Trustees to implement the terms of Henry Lawton Blanchard's will. Richard tried a new tack. Why not restore the original tavern, formerly in the Blanchard family, and operate it as a tavern of the 18th century? This time the obstacles melted away particularly when one of the trustees passed away and a replacement had to be voted by the remaining four. They bought the idea of a restored tavern with the tacit understanding that Richard would do most of the work. Richard agreed and was voted a trustee.

In the next few years he started the research to provide him with a reliable image of a colonial tavern. Williamsburg was the only place where Richard could find examples of a functioning tavern and he examined the Raleigh and Chowning taverns. He also devoured several books that provided snippets of relevant information about tavern life, what they drank, what they ate, how they amused themselves, etc. The Williamsburg taverns had strolling minstrels but they could barely be heard over the din of people eating their meals, ordering their food, and busboys removing dirty dishes. In his mind Richard decided that camaraderie should be a prominent ingredient in creating a tavern 'atmosphere' and distractions were to be kept to a minimum.

A major problem faced the trustees. The old Blanchard Tavern had been converted into a town hall and library in 1939 so that negotiations had to be initiated to purchase the building. The town was actually happy to receive this offer since it had felt cramped in the old building. A new town hall would be welcome. The trustees bought the historic building in 1975 and Richard began his career in restoring and eventually running a colonial tavern.

A postgraduate student of history was engaged to research the history of the building and the Blanchard family, and to provide whatever help she could to guide the trustees in their restoration efforts. She traced the building back to its first sale in 1752 suggesting a probable date of construction sometime in the 1740's. Richard's discussion with 'experts' at Williamsburg,

Sturbridge Village, the Newport Restoration Society and anyone who claimed knowledge of the past indicated that they all knew the intimate details of colonial taverns. What a pity that no two of them agreed!

Common sense and compromise became the guiding principles during the restoration. Fortunately one of the trustees headed a local construction company and he in turn was able to engage and supervise local workmen. The first task involved the removal of all superfluous modern additions such as floor and ceiling tiles and wallboards. The original wide pine floors were found intact under the floor tiles as well as some of the original wainscoting and ceiling beams in a structurally sound building. About twenty tons of tiles and wall covering were removed. All paint was stripped, the original fireplaces repaired, and the original doors stored in the attic were rehung. Replica wall sconces were installed for lighting and an original bar found in a tavern in Wareham, MA was duplicated for the taproom.

Physics came into play to help in planning operations. An article by Professor Sanborn Brown on colonial drinks came to Richard's attention. Brown had spent 30 years in researching colonial drink recipes employing his students as guinea pigs. Having never lost a student he produced a book *WINES AND BEERS OF NEW ENGLAND*. As dean of physics at MIT he approached the subject in a scholarly fashion and produced a well-written treatise on the subject. Richard journeyed to New Hampshire to discuss details with the now retired Brown and this helped Richard decide on an 'authentic' drink menu.

Even the tea presented a research problem. The brew dumped into Boston harbor in December 1773 was a type called Bohea, named after a hill in China. It was no longer available and no one knew what commercially available tea it resembled — except Mr. Twining, director of the famous London establishment. He advised Richard that Chinese Keemun was a good modern approximation. Unlike Ceylon tea, Keemun could be brewed for hours and this was kept ready to serve in an electric urn.

The snacks, such as pork pies, meat pies, syllabub, a spice cake, etc. were selected as typical of the period. As in English pubs the patrons ordered and picked up their drinks and snacks at the bar enabling the entertainers to perform with minimal disturbance.

The menu represented a successful effort at pleasing the palates of most patrons.

Pressures from the world of physics dictated that the Sagamore Conferences needed a respectable patron rather than the ad hoc arrangement during its first decade. And so it came to pass that Richard and Erwin (Bertaut) representing themselves as chairmen respectively of Sagamores I, II and III petitioned the International Union of Crystallography to add to their existing commissions one on charge, spin, and momentum density. The application was submitted in 1973:

"APPLICATION TO THE INTERNATIONAL UNION OF CRYSTALLOGRAPHY

The undersigned in their capacity as organizers of the Sagamore Conferences on electron charge, spin, and momentum density apply to the governing board of the International Union of Crystallography for the creation of a special commission on electron distributions effective January 1, 1974, thus absorbing the role and functions of the Sagamore conferences. The subject matter of electron distributions (charge, spin, and momentum density) is so fundamental to the creation of a special commission it will do much to bring solid state physicists, theoretical physicists, chemists, and even nuclear physicists (positron annihilation) into closer contact with crystallographers. The Sagamore conference register now contains over 400 scientists throughout the world and its rolls have increased with each of the triennial Sagamore conferences."

This was followed with a brief history of the conferences, a paragraph on how this commission would help crystallographers, and a few paragraphs about charge, spin, and momentum density. It succeeded in swaying the IUCr and the Sagamore 'gang' were sanctified. The original dozen commission members represented USA, France, Australia, Germany, Russia, Finland, Japan, and England. What began at an Army lab in Watertown had now invaded the world, demonstrating that the research report was mightier than the sword.

In 1974 Richard was off to Australia again just 15 years after his prior visit when the site had just been selected for the Sydney Opera House. It took that decade and a half to complete one of the most impressive performance centers in the world. The occasion for the visit was the IUCr conference in Melbourne. And the Australians made a magnificent effort to welcome the attendees. A formal invitation read:

THE GOVERNMENT OF VICTORIA
Invites
Dr. R. J. Weiss
to a reception
at Parliament House, Melbourne
on Tuesday, 20th August, 1974 at 6 p.m.

Two days later the conference banquet was held in the GREAT HALL of the National Gallery of Victoria, an impressive open space some 100' × 50' with a stained glass ceiling about 80' high. The menu of pâté, poached salmon, rib of beef with vegetables, orange wedges in cointreau, and Australian cheeses was complemented with a selection of Australian wines including a Riesling, a Moselle, two clarets, and a Madeira.

Richard considered the art museum in Melbourne as probably the finest in the southern hemisphere. The edifice was partially financed by one of the pharmaceutical companies but was coming under controversy for its recent £1 million acquisition of Ben Pollack's huge painting BLUE POLES. Richard spent half an hour eavesdropping on museumgoers who were questioning the sanity of making such an expenditure on abstract art, no matter how big. The Director of the museum had correctly justified its purchase as an investment. To most Australians, betting on a horse was a justifiable investment since one didn't have to wait long for the result. But a painting of such abstract unidentifiable images seemed more like lunacy.

About this time Bob Street, a displaced British theoretician who had published some papers on antiferromagnetism in chromium, was appointed Vice Chancellor to one of the Australian Universities and Richard coerced one of the secretaries at the Australian Academy of Sciences to send him a scroll with the following inscription:

> AUSTRALIAN ACADEMY OF SCIENCES
> semper antiferromagnetismus cromiumsum
> The Australian Academy of Sciences and
> Dick Weiss
> Congratulate you on the occasion of your
> new post.
> GOD SAVE THE QUEEN
> s/ W.A.ter Town

Bob Street never acknowledged the award.

Bill Reed and Richard enjoyed a sail around Sydney harbor, one of the magnificent sailing areas. Richard also made the rounds of the opal establishments. These gems whose color varies as the light scatters at different angles was the last of the gemstones to be artificially produced. There is a regular array of inclusions in the opal which diffract light at specific angles just as atoms in a crystal diffract x-rays.

Richard returned from Australia eager to follow the progress of the tavern and to continue work on electron distributions. Sagamore V (Sagafinnsk) was beginning to require some of his attention but this was diverted for a time while Richard became engaged in writing a textbook.

Leonid Azaroff head of the Metallurgy Department at the University of Connecticut in Storrs, persuaded the Department of Defense to put up money to produce a book on x-ray diffraction. He convinced them that no book existed to meet the needs of the

70's and beyond. Leonid collared five other experts to produce a tome that would encompass the entire field. Roy Kaplow of MIT, Noria Kato from Nagoya University, Arthur Wilson from the University of Birmingham in England, and Ray Young from Georgia Tech, joined Weiss and Azaroff in this endeavor. Richard had reservations that a multi-authored book would be no more than a collection of unrelated individual contributions. But Leonid was persuasive and the team gathered at such spots as Cape Cod, Grenoble and Storrs to hammer out the problems in producing a coherent effort.

Leonid proved to be a competent organizer and ran the meeting with the right amount of control and encouragement. The final product dispelled any doubts Richard held as to the book's efficacy. It was hard work for all but the sense of purpose set the team into a spirituous endeavor. Richard's memories included numerous efforts to get Kato to tell a funny story, one of the distinct differences he found between Western and Eastern culture. Kato claimed it was impossible to produce one for a Western audience yet he insisted there were Japanese comedians. It was certainly true that the Japanese derived pleasure watching TV programs where individuals are embarrassed, frightened and otherwise incommoded, situations westerners would consider in bad taste.

The book was published by McGraw Hill in 1974 but it only enjoyed a limited success. Its 650 pages, including 30 pages of references, covered most aspects of the interaction of x-rays with matter.

Richard's retirement from Watertown and his going away party followed standard recipes. He did make one parting request of the administration when he asked to be taken on as a non-paid consultant. This would provide him with a pass to visit the arsenal whenever he wished. He was turned down and never discovered whose toe he had stomped on within the upper reaches of the administration to deny him this privilege. No matter — he was getting ready for his new career. Actually, his retirement was just in time — it coincided with the first signs of memory failure.

He ran into Bill Lipscomb from Harvard who had been awarded the Nobel Prize only a few years prior. Bill introduced Richard to his friend but Richard could not reciprocate — he had forgotten Bill's name!

WATERTOWN 6: 1976–1980

The Sagamore register of 1975 contained over 400 names. Malcolm Cooper, soon to be secretary to the IUCr commission on charge, spin, and momentum density, had finished his thesis on Compton scattering and moved from Cavendish Lab to the new University of Warwick, near Coventry. Impatient to wait for Sagamore V in 1976 and anxious to show off the new University, Malcolm announced a summer school on electron density and lured about 40 to his university in the Midlands, not far from Stratford. His wife had produced twins which he referred to as a well-resolved doublet, a term employed in spectroscopy for two closely spaced but distinguishable lines.

Looking at a group photo of the smiling faces at the Warwick gathering, including eight beards, Richard was struck with the sense of purpose amongst the attendees. The fields of momentum density measured with x-rays and magnetic spin density measured with neutrons had been mushrooming and the theoreticians were now actively trying to calculate these quantities. Thus, the theoreticians did not have to seek their own corner at the conference to converse amongst themselves — they were equally comfortable listening to experimentalists describe their results and difficulties.

For some mystifying reason the architect who designed the University had selected weather-resistant white tiles to cover the exteriors of many of the buildings giving it a loo-like appearance. But the school was fortunate in that a dowager American benefactor had anonymously donated oodles of money to build a performing arts center. Their back stage set-construction facilities rivaled the best in the country. Even Stratford had some sets fabricated at Warwick.

Of course Cooper had arranged a day at Stratford so that the foreigners at the conference could try to decipher

Shakespearean English. If the Schroedinger equation didn't stump them why should the Bard of Avon? The Sagamore spirit of relaxation and informal scientific dialogue had now become modus operandi.

Richard moved on to Holland. Very little about electron distributions was presented at the huge IUCr convention in Amsterdam affording him ample time to explore the Dutch capital. Amsterdam was an ideal city to negotiate on foot or by canal boat. Richard was impressed with the Moses and Aaron Church, now desanctified and turned over as a sanctuary for anyone who needed a roof over his head. People lived inside, set up shop selling their trinkets or any other profit-making commodity, gave soapbox speeches and did whatever took their fancy. It was indeed a haven for the indigent trying to make some money and may have been the historic reason for Jesus throwing the money merchants out of the temple.

This was the first meeting of the IUCr since the Sagamore conferences had come under its aegis. Having served two terms as chairman Richard thought it was an appropriate time to follow George Washington's advice and set a precedent by resigning after this period. Bertaut became the logical successor with Malcolm Cooper acting as secretary to keep things in order since Bertaut shunned some of the communication devices of the 20th century. He refused to install a telephone in his flat — when needed one sent a messenger.

After returning to Watertown Richard directed his attention to a subject virtually ignored by the physicists, i.e. polymers. These substances occurred in nature such as fibers in wood, hair, silk, cotton, fingernails, etc. although inorganic man-made polymers like polyethylene had been discovered in the 30's. Polyethylene consisted of a saw tooth chain of carbon atoms with a pair of hydrogen atoms bonded to each carbon. It was the simplest polymer to visualize over a short distance although over long distances it bent in a tortuous fashion. Each of these chains were bonded to neighboring chains by very weak electronic forces called van der Waals forces. Ordinarily a piece of polyethylene looked like a bowl of cooked spaghetti when viewed on a molecular scale although it was possible to stretch it and straighten out the chains. Paraffin was a variant of polyethylene in which the chains were fairly short. To a physicist accustomed to neat crystalline arrangements of atoms and alloys polyethylene and paraffin appeared too messy for them to deal with, hence their lack of interest.

This tangled arrangement of atoms made it impossible to measure the electron positions around the carbon and hydrogen atoms. However, the Compton profile measurements were not so encumbered. As a result the theoretical momentum distribution calculations of Duncanson and Coulson in the 1940's were shown to be much too simple while the more recent calculations by Epstein at Brandeis were found to be in good agreement. This was the first time an experimental check was made on the electron distribution of an inorganic molecule.

The chemists had developed a simple picture for a substance like polyethylene and paraffin. Each carbon atom had four bonding electrons and was surrounded by four near neighbors, two carbons and two hydrogens. Since electrons bond in pairs the four electrons were allocated one to each neighbor and were called C-C and C-H bonds respectively. Employing this simple picture within the framework of quantum mechanics the electron momentum distributions could be calculated to as high a sophistication as time and patience permitted, just as Hartree had done for the electron positions on isolated atoms. While the Duncanson and Coulson effort of the 1940's had been rather primitive Epstein had the advantage in the 1970's of faster calculators. All in all, Epstein's good agreement with the experiment verified the simplification of separate C-C and C-H bonds, a further triumph for the Schroedinger equation, chemical intuition, and the Compton profile measurements. In later years Richard found small departures from this approximation but this was progress!

Having verified the electron momentum distribution in polyethylene Richard departed from his career of chasing electrons and became intrigued with the atomic arrangement of polyethylene, thus donning the hat of the crystallographer for the first time. No one had any trouble with the zigzag model of the carbon atoms for the backbone chains nor the arrangement of the hydrogens attached to each carbon, but how did these chains stack together? The 'bowl of cooked spaghetti' was but a crude picture that allowed for bending of the chains. Richard tried to simplify the arrangement by stretching the spaghetti into an ordered arrangement, i.e. one to resemble uncooked spaghetti out of the package. What exactly was the arrangement of the zigzag chains relative to their nearest neighbors?

Neutron diffraction measurements had suggested a model for the zigzags but when Richard repeated these with x-rays he could not verify these results in detail nor could he suggest an alternative model. In short, he had reached an impasse and decided to submit the paper for publication. It was rejected — no one liked negative results. Richard tried to induce others to repeat his measurements (shades of the 1950's when he had worked on the 3d electrons on iron) but he never succeeded. At this stage in his career Richard merely shrugged it off. It was not important — people would use polyethylene bags and not care a whit about the atomic arrangement. As long as the bags did not leak who cared?

The structure of polyethylene was not the only unsolved problem that confounded Richard in the 70's. The billiard ball collision of electron and x-ray that had led Compton to a simple formula for the energy loss of the photon only required a knowledge of the x-ray wavelength, the scattering angle, the velocity of light, the mass of the electron, and Planck's constant. Every one of these was either known to high accuracy or could be measured. Still the net effect of this fundamental process in physics turned out to yield an answer 1% too small. This was first recorded by DuMond and Kirkpatrick in 1937 but experimental techniques had improved over 40 years. Both Richard at

Watertown and Cooper and Holt at the University of Warwick made identical determinations and published a joint paper, calling it the Compton Defect.

They could proffer a qualitative argument as to the reason but when the results were turned over to the well-known theoretician and future Nobelist Roy Glauber at Harvard, he was unable to find theoretical arguments to support the presence of the Defect.

"I worked quite hard on that problem and wasted a lot of time," he later complained to Richard.

But Glauber would not publish his unsuccessful theoretical effort. That is the nature of the physicists' psyche that had emerged from their socio-scientific interactions.

1% was a big effect when the experimental uncertainty was only 0.2%. The mystery may be waiting for someone with a pipeline to the Deity to unravel it. One would have thought that theoreticians would have swarmed over such a fundamental problem. The Compton Defect heads the list of questions that Richard has prepared for St. Peter.

Sagamore V (Sagafinnsk) was held in a conference center about 15 miles north of Helsinki in a sylvan setting that matched the original Sagamore site in New York State. Professor Kaarle Kurki-Suonio at the University of Helsinki assumed the administrative duties with a result that pleased everyone including the 10 Russian attendees who were happy to savor the Western atmosphere. Eighty participants representing 15 countries congregated in Finland to report on their experiences with nature's fundamental particle the electron. It was almost a century since J.J. Thomson had discovered the little bugger, and over ten years since the Sagamore conferences had begun reporting in detail on what the creature was up to.

Helsinki was well off the beaten path and was not inundated with tourists even in the summer. Richard felt it was well worth the extra time getting there. The view from the University of Helsinki on Siltavuorenpenger Weg took in the harbor, its islands, and the outdoor market at the waterfront. One of the islands Suomenlinna boasted an old fortress well worth a day's excursion. The countryside was exhilarating and it has defied Richard's ability to rationalize this with the perpetual display of moroseness for which the Finns were noted. Perhaps the long dark and cold winters have manifested themselves in the Finnish genes. Alcohol was the mother's milk for these inhabitants of Helsinki, a city called 'The Daughter of the Baltic'.

The story made the rounds of the Finn and Swede silently drinking at a bar. With each round the Swede lifted his glass and toasted "skol" while the Finn remained silent. After the fourth round the Finn finally responded,

"Look — did you come to talk or to drink?"

Timo Paakkari, who had spent a year in America, claimed that Americans drank more than Finns. The difference was that the average Finn imbibed it all on Saturday night and spent the next day sleeping it off. The paddy wagon made the rounds of Helsinki on Sundays and collected those who had fallen into a stupor in a public park. The sotted Finn went along quietly certain in the knowledge that he would be fined 5 or 10 Finmarks. One Sunday morning Richard was sitting on a park bench overlooking the harbor and waiting to board the ferry. The warmth of the sun took its toll and he was soon in the arms of Morpheus only to be rudely awakened by a policeman's club on his knee. The gendarme's display of anger and menacing language was only assuaged with Richard's retort in English,

"I'm waiting for the ferry and must have fallen asleep."

The constable moved off, dismissing Richard with a wave of the hand. Dozing on a park bench was not illegal on a Sunday morning provided you were American.

One of the daily rituals was the sauna but there was more to this tradition than a mere steam bath. Kurki-Suonio introduced Richard to this ceremony at his family's lakeside cabin. Kaarle indicated that there were enough lakes in Finland for each native to have his own. Richard tacitly followed Kaarle about while he prepared the sauna (pronounced sow-na) realizing that questions about tradition were meaningless.

Early in the day a wood fire was started under the large boulders that were heated to retain the necessary calories needed for the entire afternoon. Such boulders had been especially selected for their ability to retain the heat, no doubt an optimal combination of their high specific heat to maximize energy storage and their low thermal conductivity to minimize heat loss. They would remain in the family's possession for decades. Next was an excursion into the woods to cut birch branches in four-foot lengths. Just before the sauna was to be used a measured amount of water was poured over the rocks to fill the room with steam.

Soft pine benches, held together with wooden pegs, provided seating for the bare occupants. Metal nails could not be used for they would transmit too much heat to one's derriere and cause a burn. The temperature in a sauna could reach over 100°C provided the air was kept dry. How could the body withstand a temperature as high as the boiling point of water? It couldn't, but the moisture that evaporated from the sweat glands removed sufficient heat from the body to leave the skin at a much lower temperature.

One sat in the sauna for some twenty minutes, employing the birch branches to flagellate one's back. Richard did not find this treatment particularly exhilarating but he had nary any Finnish blood in his veins so he couldn't claim his personal observation as significant. Not surprisingly, Finns maintained a silence in the sauna although comments by Richard were tolerated. After this twenty-minute period one dived into the cold lake and remained as long as the body could tolerate the 10°C water temperature. Richard survived this plunge for less than two minutes and returned to the sauna.

The periodic exposure to the temperature extremes of sauna and lake lasted for three or four cycles before one rang down the curtain, showered, dressed, and drank some non-alcoholic beverage. And there you were, refreshed and ready to go! Was this habit forming? Did it alter Richard's psyche? Was he prepared to install one when he returned home? The answer was no.

Kurki-Suonio assigned seven speakers to give review talks of the various theoretical and experimental facts of electron distributions. Supplementing these reviews were 56 short papers on individual contributions. From the outstanding level of the scientists involved it was clear that the subject was receiving the kind of attention that would clarify the fine details of electron behavior.

At the conference banquet Bertaut reiterated his memory of the early Sagamore conferences in New York State and the possible encounter with a bear suggesting the following bit of doggerel;

THE BEAR AT SAGAMORE V
On a hot Finnish eve dreamed Bertaut
Unclothed, from his sauna dids't go,
He was calling for beer
Instead he got bear
Now he's bare in his bier, but not chaud.

The unanswered question was: What was an American bear doing in a Frenchman's Finnish dream?

The best suggestion was that there were no national boundaries at a Sagamore Conference. In an English advertisement in a Finnish tourist magazine Richard found:

THE POET'S WAY RESTAURANTS
Licensed to sell bear and wine

Following the conference Richard was invited to spend two weeks in the university's physics department with accommodations provided by the school. Richard also planned to visit Turku and to do some further sightseeing.

At the end of this fortnight Kurki-Suonio held a physics department party at Paakkari's home. Richard had brought along a few bottles of 100 proof bourbon that he had purchased at the Commissary in London. Liquor was prohibitively expensive in Finland. To this offering the host had added some chilled aquavit and chocolates, a combination Richard found pleasant, indeed. For entertainment Richard borrowed a guitar and induced a member of the department to sing some bawdy 18th century songs that he had collected in his tavern research. This display of camaraderie underscored the fact that 'electrons held people together'.

Perhaps the event that demonstrated how very civilized the Finns were occurred while Richard was walking along a well-trafficked street in downtown Helsinki. A loud bang announced a crash between two vehicles. Both drivers pulled to the side of the road, emerged from their vehicles and rapidly approached each other as if to engage in fisticuffs. Instead, they cordially shook hands and exchanged calling cards, never raising their voices.

Finnish architecture impressed Richard. The Temppeliaukio church was hewn out of rock below ground level and was covered with a flying saucer-like roof. Finlandia Hall, Helsinki's Concert and Conference center stood by a small lake and provided the city with a large aesthetically pleasing venue for these functions. The Sibelius monument, a panoply of metallic tubes to simulate organ pipes and welded together to remind one of a tree, was displayed in a park of its own. Marimekko fabrics, jewelry to carry out archeological designs, Finnish crystal cocktail glasses carrying out a rough-hewn appearance as though chiseled out of a solid piece of glass, and tasty Finnish cheese have all made their mark in international trade.

Richard arranged for an excursion to Leningrad on the Bore Line, a three-day trip that did not require a visa. This provided a convenient arrangement that enabled the Russians to pick up some hard currency and the Finns to drink themselves stinko on duty-free booze. The boat left on a Friday afternoon for an overnight journey, to return two days later. Richard found only a few non-drinking passengers. The Finns, though three sheets to the wind, were all well behaved.

Upon arrival in Leningrad passengers were required to exchange their passports at the foot of the gangway for a visa-like card. An Intourist guide met the group and cautioned them to follow her at all times. They walked on the Nevsky Prospekt, Leningrad's Broadway, visited several churches including St. Isaac's Cathedral, the huge Palace Square, one of the parks, and lastly, the Hermitage, before being let loose to spend their hard currency in a designated gift shop.

Without doubt the interior of the Hermitage and its ornate Baroque-style salons, hundreds of them, made it the art museum of art museums. The paintings were virtually overshadowed by the interior décor.

In the 162 page February 1977 issue of the Scandinavian Journal of Physics PHYSICA SCRIPTA the preface by Kurki-Suonio traced the modus vivendi of the Sagamore Conferences and the rationale for devoting the entire issue of the journal to a review of the progress since 1964. In seven well-edited articles by Bruce Forsyth, Art Freeman, Brian Williams, P. Lindner, Pierre Becker, Roberto Colella, and Vedene Smith, Jr. the entire compass of theory and experiment in charge, spin, and momentum density was reviewed by the experts. In the theoretical paper by Smith the electron density in water was shown in calculations to high accuracy. The bear that we all thought was being confused with beer was actually hiding in water.

In the same year (1977) a McGraw-Hill book edited by Brian Williams traced the history and achievements of Compton Scattering in measuring momentum density. With all the worldwide attention devoted to electron behavior it was obvious to Richard that if Slater were still around he would have been pleased with the progress. Of course, it was an unending process since neither theory nor experiment could be performed without a few percent uncertainty.

Upon his return to Watertown it had become apparent that Richard would have to wind down his quarter century of flirting with the elusive electron. Basic research had outpaced Federal funding and the Army had decided that all work had to be relevant to its mission. A new Director Al Gorum had already entered the scene, one who Richard approved of and was partly instrumental in his selection. As long as the tide had turned and the longhaired physicists had to direct their attention to applied physics it was pleasant to have someone who could speak their language. Gorum agreed that any continuation in pursuing the electron would have to be accomplished 'under the table'. First and foremost he had to demonstrate to Washington that the Watertown Laboratories could help the Army. Richard had to find a practical avenue of research with a solid end product. He turned his interest to his former fascination with thermodynamics.

John Mescall, one of the mathematicians at Watertown, had been making calculations of the penetration of projectiles through armor. It had always been assumed that tough metals made the best armor and a considerable amount of ballistic data was generated on firing ranges. In one test a brittle but extremely hard ceramic Al_2O_3 (sapphire) was tested and the results surprised everyone. Even though the bullet shattered the ceramic the bullet was also destroyed (comminuted was the word coined for such disintegration). What caused this? As Mescall pointed out the harder the armor the higher the pressure in the shock wave generated in both the armor and the projectile. Before the forward-going shock wave reaches the back surface of the armor the return shock wave hits the back surface of the bullet and pulls it apart in tension. Most materials are stronger in compression than tension.

With this result a new philosophy developed in fabricating armor. The surface absorbing the projectile's shock had to be very hard, even if brittle, and this should be backed with a tough metal to prevent the fragments of the hard material from penetrating the armor. In addition to this strategy it was obviously desirable that the armor be lightweight and not excessively expensive. Richard and his colleague Ken Tauer were aware of the development of titanium diboride TiB_2 as a very hard ceramic that was 50% lighter than steel. While the powdered raw material was not too expensive it required high temperatures and some pressure to produce pieces that were fully dense (all holes removed) adding to the cost.

Serendipity came into play over a hunch that heating a mixture of titanium and titanium diboride powders would react and produce a new boride. Indeed an exothermic reaction produced the following:

$$Ti + TiB_2 = 2TiB$$

The product titanium monoboride was almost as hard as the diboride but could now be made at much lower temperatures. Large samples were made at a private company in Cambridge and these were backed with titanium to provide a shock absorbing substrate. Ballistic tests proved this new lighter weight armor to be worth further development but as often occurs in the military the funds ran out — the country was not at war and armor was no longer a high priority item.

Pierre Becker in France must have been pleased with Malcolm Cooper's summer school at the University of Warwick for he was motivated to do even better two summers later. He selected Arles on the French Riviera, a town normally associated with Van Gogh, and engaged a reasonably priced hotel (Primotel). Invitations were sent to those on the Sagamore register. When Richard received his he had to find an excuse to inveigle the Army to pay his way. He submitted to Washington that he had been invited to lecture at a NATO summer school. Becker knew someone at NATO headquarters and arranged for Richard to receive an invitation on NATO stationery. Chicanery with a lofty rationale.

With women and children included, the group photo contained 125 participants, Pierre forcing a grin in the front row that did not disguise his concern over a fight he just had with the manager who for the fourth day in a row served the identical dessert at dinner. Absent from the photo was Professor Daudel, an apparently stern taskmaster who arranged for a friend Madame D'Agaggio to exhibit her painting at the hotel. The young Pierre seemed to move in dread of Daudel who kept to himself. Richard, a contemporary of Daudel, was intrigued and with a little effort he broke down the barriers for communication as they swapped war stories. Daudel had been an undercover agent working as a gendarme in a Paris police station. When disabled V-1 rockets fell in France members of the underground would deliver the pieces to the back entrance of the police station where Daudel managed to reconstruct a complete rocket for shipment to England by fishing boat. What better cover than a Paris prefecture to hide this operation! Madame D'Agaggio's paintings were most professional and Richard was sorry his budget could not afford one.

Daudels' scientific work focused on solving the structure of a habit-forming drug molecule and then trying to synthesize a similar but harmless molecule that would fool the body but be devoid of withdrawal symptoms. A lofty effort!

Pierre exerted every effort to please the attendees during their afternoon excursions through the town and to interesting sights nearby. The town itself still kept its myriad of narrow and charming 17th century streets and Richard was amused when a Cadillac inched down one rue only to find it could not execute the turn at the corner. By this time other cars were lined up behind the Cadillac and honking their horns. A few cars were also attempting to proceed in the opposite direction (no one way streets) and they all met at the Cadillac. It was all too humorous but Richard did not hang around for the unsnarling — he had other things to do.

Two coliseums, one in Arles and one in Nimes were in an excellent state of preservation, far better than the one in Rome. Most impressive was the aqueduct at Nimes, which carried water from the hills into the Roman settlement at Nimes. The slight fall in elevation had to be just right for if the gradient was too steep the water would not reach the city. The aqueduct was lined with lead to prevent leakage and it has been postulated that the fall of Rome was due to lead poisoning. Even today the gradient is still accurate and the monument to Roman engineering is there for all to behold.

One unforgettable incident occurred while Richard was accompanying a female participant on the streets of Arles. A Mercedes slowed down and an outstretched hand grabbed the lady's handbag. Richard pursued the culprits on foot but the Mercedes had too much horsepower. In all the years Richard had lived in crime-ridden America this was the first time he had witnessed a 'violent' offense.

A side trip to Avignon and the Pope's Palace, an impressive edifice used by the Pope's from 1305–1376 when Rome was in turmoil, provided an interesting interlude although not as fanciful as the famous bridge that inspired the song 'Sur le pont d'Avignon' on which people were said to dance. As the guide pointed out, the bridge was much too narrow for dancing, such activity was carried on <u>under</u> the bridge. The song should be sung 'Sous le pont d'Avignon.' After a journey to the Camargue, an area set aside for wild small ponies, the conferees shared a picnic at the beach. Walking along the Mediterranean a participant from India joined Richard to complain of the immoral behavior associated with displays of nudity on the beach, although that did not deter him from ogling the pulchritude.

The conference was a success in spite of the lack of variety for dessert at the hotel. In reporting on events the Arles newspaper publicized the conference but in much larger headlines described the paintings of Madame d'Agaggio!

In another excursion Jane Brown, a former colleague at the Cavendish Lab, and two others visited some of the wineries. At one time Jane was official wine taster at her college and she was shopping for Chateauneuf du Pape. She purchased many cases to be laid down for several years before tapping. Richard was impressed at her discrimination since he could scarcely taste the difference from one vintner to another.

In a talk at the conference Richard resorted to cartography to review the problems in correlating charge, spin, and momentum density as solutions to Schroedinger's Equation.

He called the chart 'Le monde de l'electron' Central to the map is La mer Schroedinger (the Schroedinger sea) which had to be traversed to reach the various countries. In the west there is 'le pays r' (country of position) where one finds the electron inhabitants at various positions but one could never see them furtively steal from one place to another. When Richard began his career he spent the bulk of his time in le pays r but achieved limited success. His x-rays could not always penetrate the ever-present jungles.

When he later began his cooperation with Cooper, Leake, and Phillips, Richard moved eastward to 'le pays p' (country of momentum) where the inhabitants are always flitting about but could never be found to come to rest. Safe passage from east to west and the return journey depended upon finding the quantum mechanical wave equation solution to the Schroedinger Equation. There is 'le peninsule K.E.' (the kinetic energy peninsula) which can be directly evaluated from the Compton profile.

North of the Schroedinger sea is le port E (the total energy port), one of the easiest properties to calculate. It is a safe haven, made up of one part positive kinetic energy and two parts of negative potential energy. This was the first property measured in pre-quantum mechanics days. Another peninsula that can be reached from measurements of the charge density in le pays r is the potential energy V.

Stormy seas with perennial hurricanes were present so that it was impossible to proceed from le pays p to le port E without crossing la mer Schroedinger. Lastly was the spin density for both position and momentum, the former measured with neutron diffraction, the latter with circularly polarized x-rays.

Exploration of the various countries had been historically random and uncertain but at least the map was there for all to use when entering le monde de l'electron. What fun! In two 1978 publications Richard had given the first indication of how to cross into the peninsula V and following a suggestion by Epstein how to cross the narrow inlet into peninsula K.E. These two routes have remained untrodden.

With this tongue in cheek exploration into the map of the electron's world Richard had left several guidelines for future settlements. It turned out to be his swan song for he now looked forward to becoming a tavern keeper. No one who would be entering Blanchard's Tavern would be expected to be looking for electrons. He would leave it to the younger scientists to capture the swans sailing the Schroedinger Sea. Hopefully, a cookbook for Le Cygne de le Mer Schroedinger would be published.

By 1979 the tavern was nearing completion and Richard began to line up drinks, snacks, and entertainment. A mime, singers, and, storytellers became part of the crew of the entertainers. Richard, himself, turned to writing and produced a series of satirical letters that he planned to read to the patrons.

He also obtained xerox copies of the Boston Gazette for the year 1780, a newspaper that rivaled modern papers in that most of it was devoted to advertising. A sample of one of Richard's satirical letters purported to be from General Washington:

CONTINENTAL ARMY HEADQUARTERS
Cambridge, Mass^{tts} Bay Colony

Blanchard's Tavern
East Stoughton, Mass^{tts} Bay Colony
Jan 10, 1776

Gentlemen:

It has come to my attention that a rumour is being circulated about myself and your establishment in East Stoughton. I beg to inform you that at no time have I ever slept in Blanchard's Tavern. However, last fall on a journey from Boston to Newport I distinctly recall my horse stopping under the large oak tree in front of Blanchard's Tavern, and he did appear to have been well relieved as a result for he renewed the journey with greater vigour.

You are therefore entitled to mount a brass plaque on that tree (which must surely have profited from that visit) stating that General Washington's horse stopped here — or any phrase that you feel is more descriptive of that event.

I remain yr ob servant
G. Washington
General, Continental Army

Sagamore VI would bring the venue back across the Atlantic to Mt. Tremblant in the Laurentians north of Montreal. Vedene Smith at Queen's University in Kingston assumed the burden for the local arrangements. Richard decided that he would drive up from Watertown and communicated with several overseas visitors arriving in Boston to ask them to join him on the journey.

With the conference half a year away Richard turned his attention to a military problem that had haunted aircraft and naval designers for years — turbulence.

Heisenberg was reputed to have said that this was the most difficult problem he had encountered in physics. Any object moving through air or water at high speed encounters frictional forces between gaseous or liquid molecules that touch the object as it moves. In bouncing off the object these molecules collide with other gas or water molecules and eventually create a disorganized flow. This rampant movement builds up suddenly and leads to a complex eddy-like flow called turbulence. An airplane or a ship that encounters turbulence (and they all do) must expend 10 to 30% more fuel to drive the vehicle.

Since the problem defied theoretical solutions, pragmatic efforts had been made following hunches generated by scientists. In 1964 Northrop Aviation revamped the wings of an aircraft by drilling slots along their length and installing vacuum pumps inside the wings. It actually worked and eliminated turbulence. However, the Vietnam conflict terminated the research program and further effort was dropped. But that changed in 1973 with the Israeli War and the reduction in gasoline availability. The aircraft industry suddenly faced the possibility that fuel rationing would limit aircraft use.

Working with the Navy Labs in Newport Richard helped with the question of strong porous materials that might be employed in aircraft and ship construction. But the Department of Defense chose not to fund the project since this type of material did not have a clear military mission. (Saving fuel was not enough).

While the Sagamore Conference in the Laurentians continued the tradition of 'wooing the electron' and pleasing the conferees Richard's most fascinating memory occurred on the drive back to Watertown. In passing through Montreal he ran into a long delay since they were holding their first world marathon and one of the major bridges was closed to auto traffic. Some 20,000 contestants competed and it required some devious maneuvering over a circuitous route to get across the St. Lawrence. Imagine Richard's surprise to discover that the race was won by someone he knew who lived in Avon, a town of but 5000! The winner was, in fact, the younger brother of the tavern mime!

With the termination of the liberal research program at Watertown and the imminent opening of the tavern the lure of retirement loomed large. Richard was offered Visiting Professorships at King's College London and at the University of Surrey. He opted to try a retired lifestyle, England during the summer, California during the winter, and Avon during the spring and fall. Whatever opportunities presented themselves at these three locales he would be in a position to respond to them. Richard mulled it over and considered retirement at age 56 an attractive prospect.

ENGLAND: 1980–1990

Richard found the area of London surrounding King's College on the Strand to be the most culturally intense he had encountered in the English-speaking world. Within a fifteen minute walk one could reach the theater district, dozens of museums, the huge South Bank complex, the Barbican Centre, Covent Garden, etc. The ten years Richard eventually spent at the College were rich in historical and aesthetic discoveries and inspired him to try a number of sidelines that nurtured these interests. There was scarcely a lunch hour when Richard did not attend a talk at the National Gallery, the National Portrait Gallery, the Museum of London, the University of London, the Royal Society, etc. or attend a recital at Royal Festival Hall or at any of the many churches which provided such entertainment. Music, history, science, art — whatever elevated the human spirit was available to be enjoyed and at little or no cost.

His second appointment in the Metallurgy Department at the University of Surrey in Guildford provided less in the way of perks, it was strictly science. Richard was primarily utilized to advise the students engaged in their thesis research. The first year or two was devoted to finding his way about the two colleges and familiarizing himself with the members of staff and the research that overlapped his own interests.

Michael Hart, Wheatstone Professor and Head of Department at King's College, had arranged for Richard's Visiting Professorship. He had established a distinguished reputation in x-ray physics leading to the coveted appointment as Fellow of the Royal Society. Richard suggested to him that they try out a new technique for x-ray diffraction, the energy dispersive method. No company manufactured a commercial unit dedicated to this technique so that Richard had to find a used medical x-ray unit — one that a hospital was willing to discard. A new unit would have been prohibitively expensive. One required a high voltage x-ray generator, one in the 150,000 to 200,000 volt range. Ordinary x-ray diffraction generators were limited to 60,000 volts.

Normally, one employed a single energy beam of x-rays and examined the scattered intensity at various angles. The energy dispersive technique placed the detector at a specific scattering angle and examined the energies of the x-rays. Both techniques provided information on the crystalline arrangement of atoms but the latter had certain advantages that Richard wished to exploit. It was an idea that emerged from his fascination with thermodynamics.

In any solid the atoms are in constant vibration depending on the temperature and the forces between atoms that are governed by the electronic arrangement. All substances differed in the strength of these forces and physicists referred to the peculiar manner of the collective vibrations of the atoms as phonons, adopted from the prefix phono signifying sound and the suffix photon signifying its energy as h, Planck's constant, multiplied by its vibration frequency. The simplest example of such collective motion was a sound wave passing through a substance. It has a velocity and a wavelength. As the temperature increases more and more phonons of shorter wavelength (high frequency) are created. If one could measure the relative abundance of the phonons at various frequencies one could calculate the specific heat and from this the thermodynamic stability of the different phases such as alpha and gamma iron.

Neutron diffraction had made some inroads into this frequency determination but the technique was limited to only a few substances. It was Richard's intention to measure these vibrations from the manner in which they affected the x-ray scattering. A used medical x-ray unit was turned over to the physics group by the King's College Hospital and installed in a basement laboratory. It ran a few hours before the tube burned out. That spelled finis to the effort, at least until some sort of replacement could be found. With time on his hands Richard sought other fields to conquer. He returned to Massachusetts in the fall.

Back in Avon the Tavern opened with little fanfare. Richard had a colonial costume fabricated by an historian who specialized in period tailoring. His green wool tavern keeper's costume consisted of breeches and a jacket with pewter buttons offset by long white stockings that covered the knee. These were easily obtained although the salesperson looked askance at Richard. A white linen shirt that fell to below the knees and could double as a nightgown was procured from either of two commercial mail-order vendors. When worn with breeches the lower reaches of the shirt were rolled up and tucked in causing the derriere to be exaggerated. By this time in his life Richard began wearing reading glasses and he discovered that a pair found in the basement during an earlier restoration effort was ideal in both style and lens power. Lastly, what about a wig?

Wigs went out of favor at the time of the Revolution since they were associated with English nobility. Still, Richard felt that such an item of attire might elevate the tavern keeper's status in the make-believe 18th century community. He had the bright idea of purchasing a used barrister's wig in London and on his next visit to England he found a shop catering to the needs of the legal profession on Chancery Lane. Richard boldly approached the salesperson,

"Do you have any used barrister's wigs?"

"I'm afraid not. We are always ready to buy these ourselves since they were fabricated of the finest horsehair."

Richard registered surprise, "Well, how much is a new wig?"

"Approximately £200."

Richard was struck dumb and the clerk was quick to explain, "They are pure horsehair, sir."

"I suppose the horses are pedigreed."

The clerk smiled but proposed no path out of this difficulty.

"What about theatrical wigs?" asked Richard.

This struck home. "Ah, yes. There is a shop off Trafalgar Square. It's called the Theatre Zoo."

Richard trotted on down and bought two eighteenth century nylon wigs for £9 each. They have since served the tavern keeper quite nicely, protecting his bare pate on cold evenings.

Thusly dressed Richard was ready to greet and entertain patrons that drifted in from Avon and surrounding communities. To Richard Blanchard's Tavern was the equivalent of a local pub, a place for socializing over a drink. It removed from his life the uncertainty as to where he should go come Friday and Saturday night. Surrounded by imbibing patrons he never drank while serving as tavern keeper. It reminded Richard of one of his first part time jobs in an ice cream parlor. He never touched the stuff.

Sagamore VII was scheduled for August 1972 in Japan. Having left the employ of his benefactor at Watertown Richard had to rely on the generosity of the Japanese and they were munificent in providing funds. Yet in the deep recess of his subconscious he could not totally discount his war years when the Japanese were the enemies. An experience like WWII during one's formative and impressionable years did not erase easily. As is well known one never forgot where one was on December 7, 1941. Etched in his memory was the vision of the Japanese Naval Officer he observed as a prisoner of war. The look of arrogance was retained by Richard for decades even though their eyes met but for a few seconds. Richard's memories might very well have clouded his ability to view dispassionately a new generation of Japanese. Still, he was very fond of Kato and he hoped he could interact with others in a similar spirit of international cooperation although his thoughts might still expose those latent feelings.

On arrival in Tokyo Richard was struck with the stylistic incompatibility of Japanese architecture. Neighboring buildings had no relationship to each other, presumably a result of the high value of property in the city. Japan was a crowded nation since there were but a few habitable regions around which cities like Tokyo and Osaka had mushroomed. The vast bulk of its geography was too mountainous to sustain urbanization. Such incongruous sights as a tall modern skyscraper with a pagoda on the roof caught Richard's amused attention. The prevalence of a disorderly conglomeration of billboards, telephone wires, and what have you contrasted sharply with the neat appearance of the white-shirted men and the well-tailored western clothing worn by most women. While a noticeable fraction of women dressed in traditional kimono and sash the more practical western dress had been adopted for business. But in spite of this grotesque panoply of architectural hideousness one would find small Japanese gardens tucked away in the oddest places. Such a garden might consist of no more than a few rocks on a sand base with a small figurine or a stunted tree. Perhaps it is this overcrowding which has prompted the Japanese to make all available space aesthetically interesting.

The people, though, pleasantly conformed to a dress code that reflected meticulous care to cleanliness, conformity in style, and neatness. Furthermore they were well disciplined in their public and private behavior and, since food is expensive, the populace was not obese. There was a smattering of English in their conversation, mostly from TV ads which frequently identified their product by its English name.

From his reading Richard learned that the role of the Emperor was rather greater than he had been led to believe following the war. Hirohito was probably consulted about the decision to bomb Pearl Harbor but MacArthur decided to maintain the Emperor's benign image in order to assuage the Japanese. In spite of his keen interest it was impossible for Richard to broach this subject in Japan knowing how sensitive an issue it was.

Cars drove on the left and the taxi drivers were notorious in their kamikaze mindset as they darted through traffic. The drivers controlled the locks to the passenger doors — once on board one didn't leave until released. The policemen on traffic duty were dressed immaculately and performed their signaling in military fashion. The jinrickshaw, a form of transportation Richard remembered from his visit to China during the war, had been outlawed in 1945.

Food was another matter. Richard was unaccustomed to Japanese delicacies. He entered a department store in Osaka to be greeted by a line of smiling girls in uniform who added a bow to their words of welcome. He wandered around the basement where many merchants purveyed their food at individual stalls, setting out on the counter a basket with samples. Richard was forewarned with the admonition:

'If it doesn't move try it, if it does move wait for it to stop'. Spotting a basket filled with circular, amber colored rings he sampled one only to discover it to be a rubber band! One clerk offered Richard a snippet of what he called 'peanuts butter', grammatically correct but jarring on the ear.

When eating in restaurants Richard liked the idea of the painted wax-like displays in the windows that identified each dish on the menu. But at the conference particular care was take to provide western dishes although most conferees thought that they would have been game enough to try the local fare. Bob Nathans told Richard the story of his first visit to a small university town in Japan and the assurance by his host that they had found an inn where the chef had acquired an American cookbook and had practiced how to fry ham and eggs for breakfast. An hour before the appointed time for Bob's breakfast the chef cleared the decks and prepared his one American specialty. It was then placed in the icebox so as to be properly chilled when Bob entered!

Kato's charming daughter was well versed in English and informed Richard at the conference that one sauce had been prepared for all dishes! Richard had never learned to manipulate chopsticks and enquired of Kato whether the Japanese could eat a McDonald's hamburger with chopsticks.

He smiled, "If need be, yes."

The Sagamore Conference was held in the National Park at Nikko, a resort area in the hills 100 miles north of Tokyo and away from the discomfort of the city's humidity. Of the 100 or so attendees about half came from Europe and America. Every effort had been made to capture the Sagamore spirit including liberal dosages of Suntory whiskey. This represented an effort by the company to produce a drink that tasted like Scotch. At first an effort was made to distill their own whiskey but as one Scotsman described the drink,

"It was like touching your tongue to a car's engine block and holding the spark plug cable in your hand."

Suntory finally gave up and imported large quantities of Highland whiskey to blend themselves. The result was satisfactory.

At the conference dinner Richard came prepared for his after dinner talk. He had vowed to tell a funny story that would appeal to his hosts but decided he'd better meet them half way by preparing two placards in Japanese, courtesy of the flight attendant on the trip from LA.

"An American had heard about the wild taxi drivers in Tokyo and had the hotel porter prepare a sign." Richard held up the first placard, which read 'Please drive slowly'. "While the driver started off slowly he was soon caught up in his old ways and began to drive recklessly." Richard stopped to let this sink in. "On his next journey he had the hotel porter prepare a second sign in Japanese. He hailed a cab and showed the second sign to the taxi driver." Richard held up the second placard, which read 'I have high blood pressure'. Richard paused. "The taxi driver took one look and then sped him to the nearest hospital."

About half the natives laughed. Richard considered it a good start.

The Japanese had taken to baseball but not without its own stamp. While such American phrases as ball, strike, and foul were kept they did not use the term walk but called it a 'four ball'. At each game there was an incessant beating of drums and a persistent din that did not reflect the spontaneous cheering in America in response to the game's action. While Richard was watching a game on TV between the Hanshin Tigers and the Yokohama Taiyo Whales the unmentionable occurred when two Hanshin coaches beat an umpire unmercilessly, throwing him to the ground and stomping on him. The umpire was hospitalized for three weeks and the coaches were ejected from baseball for life. In an otherwise polite society accustomed to bowing and scraping this event left Richard with mouth agape — another Pearl Harbor!

At the end of the Conference Richard journeyed by the Bullet Train to Osaka. He discovered that smoking was so prevalent in Japan that only one of the ten cars was non-smoking. The well-engineered train probably achieved speeds in excess of 100 mph, passing the sacred Mt. Fujiyama, which was often shrouded, in clouds.

One of the scientists at Osaka University offered to escort Richard around the city to see some of the sights. Kakehashi appeared every morning at Richard's hotel bearing a small gift and then led him to see numerous Temples, many ornate and gilded. Soon Richard felt that if you'd seen one you'd seen them all. Yet his host could not have been more courteous. At one point Kakehashi asked Richard if he wished to use the 'gents'. Richard shook his head and immediately noted the pained expression on his host's face.

"But, if you wish to go, please do," suggested Richard.

Kakehashi beamed, bowed, and said, "Thank you," before he rushed off, returning much relieved.

One of the Palaces, former residence of some high dignitary had an interesting interlocking arrangement of the wooden floorboards such that they squeaked when trod on. This was to provide a warning if a thief approached at night. Kakehashi had not visited America and asked if there were any castles in the USA.

"No," answered Richard.

"What about the Pentagon?"

The Japanese did have a sense of humor although it was often unplanned.

At a Japanese Temple one donated money to Buddha, got on one's knees, clapped one's hands, and even left a gift of sake. This ritual prompted Richard to write the following about Kato who had spent years working on the insoluble extinction problem:

THE SPIRIT OF SAGAMORE

We entered the great Temple to pay homage to the Almighty One. And we set before him our gifts of yen and sake. And the Great Buddha said,

"Professor Kato, do you still work on extinction?"

"Yes, Almighty One."

"Then please take back the sake — you need it more than me."

Japanese science was very good although bright new ideas seemed to escape them. In fact, original thinking was discouraged in industry — conformity was de rigueur. Richard concluded that there was a rationale to this philosophy. In a crowded country conformity made life more orderly and less likely to be disrupted through conflict. Richard's vision of the country was a land with 10^8 people and 10^7 Toyotas.

At a previous conference in the UK one of the Japanese theoreticians arose to say that he doubted one of the DeMarco-Weiss measurements on the lack of spherical symmetry in the charge distribution of vanadium. Richard nodded and shrugged it off. A year later the same scientist had redone his calculations and felt compelled to rise up at a conference to publicly apologize and to admit that he now thought the measurements had been correct. Japanese honor was presumably at sake. The public apology only embarrassed Richard.

Optical engineering including optical fibers appeared to be a budding field. Richard became intrigued with this new technology and suggested to Seweryn Chomet, a colleague at King's College London, that a new journal devoted to optical fiber sensors might be welcome. Chomet proved to be a man simpatico to the published word. He had been a writer producing editorials for the London Times until he carelessly libeled someone and the ensuing lawsuit found him back walking the pavement on Fleet Street. He became a member of staff in the Physics Department and Richard found him not beholden to the rigid ideas of an Englishman. His philosophy of life was in his oft-quoted phrase,

DRESS BRITISH BUT THINK YIDDISH

A third essential ingredient in this plan to enter the publishing world was the fiber optics engineering professor Alan Rogers. Once this troika agreed to produce the International Journal of Optical Sensors a sugar daddy was required since none of the three were ready to mortgage their home on the project. Chomet had a friend who was a successful American scientific publisher and he inveigled him to put up the capital. It became a synergistic trio. Chomet knew the details of the publishing game, Richard had a materials background and spent a goodly fraction of the year in America where he could promote the journal, and Rogers was an expert in the field of optical fiber sensors and would lend respectability to the endeavor.

Richard had often tried to read editorials in scientific journals but was always bored to tears. Could he help launch a journal that made good reading? In the first issue he wrote:

A NEW JOURNAL?

There is an old saying that when two Greeks get together they open a restaurant. With the proliferation of scientific writings one is tempted to add that when two scientists get together they write a book or start a new journal.

The two editors with backgrounds in optics and in materials met in a Greek restaurant and convinced each other that photons will replace electrons for most applications where information is extracted, processed or conveyed. That conviction is the raison d'être for the INTERNATIONAL JOURNAL OF OPTICAL SENSORS.

The editors had to cajole, offer payment, collect old favors and anything else they could think of to attract contributions to a journal run by three renegades. The editorial offices were officially listed as S. Chomet, Physics Dept., King's College London while the business office had the title Newman-Hemisphere with Chomet's home address. If ever a journal started on half a shoestring this was a prime example. Fortunately the shoestring was fabricated from optical fibers and the field was expanding exponentially. With dogged determination the editors inched ahead.

They were fortunate in soliciting the help of Joe Yaver, Executive Secretary of the Optical Engineering Society (SPIE) based in Bellingham, Washington. Joe was in the process of building SPIE into the leading world organization in optics and he kindly provided Richard with membership lists and mentioned the journal in his newsletters in exchange for reduced subscription rates for members. The journal was expensive — over $25 per issue — but this was easily managed by industrial organizations or college libraries.

Richard tried a new concept in technical reporting. While interviews were often conducted by reporters who came prepared with questions for an expert and then published the interview after some editing, Richard relied on his status as a research physicist and materials scientist to engage in a genuine dialogue where both Richard and the person interviewed contributed to the unraveling of a scientific problem. The exchange was taped, transcribed, edited by both Richard and the scientist, and then published under the title SENSORED DIALOGUES. It was not conducted in a Q&A format since the dialogue rolled along once the two identified the problem. It turned out to be hard work for Richard. He found that over half of what was said during such debates was redundant, uninteresting, without literary merit, or just plain unintelligible. It did succeed in getting the story out.

Richard attended many conferences each year and collared those scientists engaged in interesting research to ask them for a paper for the journal. He would wander through the exhibit halls where manufacturers would display their new products and solicit journal submissions. Meanwhile, back at the ranch, Chomet would scream that they were behind in meeting the deadline for the next issue. What a way to enjoy one's retirement! After a few years the editors sold the journal to the London publishing firm of Taylor and Francis which promptly incorporated the journal into a new enterprise THE INTERNATIONAL JOURNAL OF OPTOELECTRONICS with Rogers, Culshaw, and Weiss as editors. But after a few years the economy took a tumble, subscriptions diminished and the three editors were eventually discharged, Richard going first.

With both his historical and scientific interests pervading his thoughts Richard turned to more serious writing. Stage plays, screenplays, monologues, novels, etc. all captured his imagination. When he met one of the scientists from Scotland Yard at an optical society meeting Richard's fondness for Sherlock Holmes sparked an idea. He created a character Robert Flowers, a Professor of Forensic Physics at King's College who consulted for Scotland Yard. He produced 12 short stories a la Conan Doyle and called the collection A DOZEN FORENSIC FLOWERS. He sent them off to an agent in New York who enjoyed them, made a few suggestions, and tried to peddle them. No one bit. Since Richard's alter ego is probably tied up in the Flower's image the first story follows:

A TOUCH OF MONET

In spite of his shock of bright red hair and an equally colorful van Dyke, Professor Robert Flowers, 5'10" and 61 years young, has maintained a low but slightly obese profile. Not since the capture of Allan Nunn May, the notorious WWII spy, has a King's College academic been so newsworthy, but as consultant to New Scotland Yard, Flowers has deemed it prudent to keep his whereabouts reasonably clandestine to avoid calls from criminal investigation divisions throughout the world. While he had a phone in his office, he deliberately excluded one from his lab — Scotland Yard paid him as a consultant and they knew how to reach him through the physics department secretary. Anonymity had its rewards, a fact he appreciated after his 55th birthday when time became more important than money. That was patently clear with the marriage of his two daughters. He gained two bedrooms, one telephone, and a goodly bank balance.

Wearing a hat to divert attention from his shock of red hair, he walked down Q corridor and entered the college's main building where he took the lift up two floors to the street level. Proceeding onto the Strand, which is bisected by St. Mary-le-Strand church, he turned left toward Charing Cross, his face cutting through the humidity of a typical overcast and chilling November day. A slight detour took him through Covent Garden, where he paused to enjoy the buskers performing in the central courtyard. Thank God he didn't have to blow his trumpet under those weather conditions. At Charing Cross he turned down Craven Street toward the Thames, passed the former residence of the first American physicist Benjamin Franklin, cut through a short alley past the Sherlock Holmes pub and crossed Northumberland Ave. to his destination at Scotland Yard. Flashing his badge to secure admittance he took the lift to the third floor and stopped briefly to pay his respects to Superintendent Frank Harney before proceeding to Michael Landon's office.

Harney beckoned him enter and Robert flopped into the old leather chair near the window. The Superintendent towered several inches over Robert, his white hair and ruddy complexion suggesting a highlander close to retirement age. The view afforded a good view of Big Ben and Robert checked the time against his Casio digital watch before he turned his attention to Harney.

"I won't keep you long, Robert, but I've been going over the file on the Osborne case — it's time we closed it out."

"Five years, two months, and three days."

"No doubt you're right. The thing that puzzles me is your final report last week that referred to the confidential Foreign Office investigation four years ago," said Harney.

"Something not clear?"

"That Foreign Office file has been locked in my safe all this time. How could you repeat seven paragraphs verbatim, including two misspelled words?"

"I can't help being blessed with a photographic memory. As to the wrong spellings, who am I to disagree with the Foreign Office?"

"Isn't there some way you can erase that information?"

Robert shook his head and pointed to his forehead, "Once it's part of my software I take it to the grave."

"Remind me to have you shot — justifiable homicide. No jury would convict me."

Flowers paused to follow the path of the graceful Concorde climbing westward, "My lips are sealed."

"I still shudder to think of the damage you could do if you talked in your sleep."

Robert smiled, "I sleep alone — most of the time."

Harney thanked Robert and showed him to the door, "How's the wife?"

"Compared to what?"

The 6'3" Superintendent playfully shoved Robert through the door,

"Off with you — and see if you can get Michael to give up smoking."

Robert proceeded along the corridor to Inspector Landon's office. Michael was in charge of the laser and high technology detection unit, a relatively new section of the criminal investigation division. His aromatic Balkan Sobranie pipe tobacco created a pleasant aura of about five yards radius. Professor Flowers frequently teased Michael that his tobacco ash would end up as a clue in a murder case someday. After greeting his visitor, Michael advised him to keep his hat on, they were heading out immediately.

Before passing the Superintendent's office Michael extinguished his pipe and then informed his boss that they were off to the National Gallery for about an hour. They descended to the street and walked toward Trafalgar Square, Michael summarizing what little he knew about a problem that had arisen about the authentication of a painting. Michael was in his early fifties, dark complected, graying hair combed straight back, and about Robert's height. He had been responsible for having Robert appointed as a Scotland Yard consultant after taking his forensic physics course at King's College seven years ago. Since then they worked closely — there was little they didn't know about each other.

They entered through the main gallery door opposite the Nelson column, and proceeded to the restoration section in the basement where they were greeted by Sir Roger van Zaal, the fifty-two year old director of the museum and well known expert on eighteenth and nineteenth century continental painters. The distinguished Director towered over the two and wore a grey lab coat and goggles to protect his eyes from the organic solvents used to clean oil paintings. He escorted the visitors into a large workroom where the paintings were strewn about in various stages of cleaning, restretching, framing and examination. On a side table in one corner rested the painting in question, a Monet from his middle period, about 1875. Sir Roger nodded at the canvas,

"This is it, gentlemen."

Van Zaal informed them it was a landscape executed in France, although nothing provided a clue as to the precise village or town. The painting was 'discovered' about three years ago in southern France and came on the market through a dealer in Arles, a town ordinarily associated with Van Gogh. However, it was well known that Monet left paintings with landlords all over France in lieu of rent. Provenance was difficult to establish.

"I assume it's been given the usual pigment analysis and x-ray fluorescence," queried Flowers.

"Yes, and four of our best people have examined the technique and brush strokes. The canvas and wooden frame appear to be original," responded Sir Roger.

"So, why are we here?" asked Flowers with a touch of waggishness.

"Four more Monets have come on the market after we bought this, all of the same period — in fact all the same scene."

Flowers smiled, "A touch of suspicion?"

"We operate on a very tight budget. If we've been duped, the Ministry tightens the purse strings, and I don't blame them."

Robert removed a magnifying glass form his pocket and took a close look, "It appears this has been cleaned?"

"Yes, we do not purchase valuable art until some preliminary restoration is completed."

"Was it dirty?"

Sir Roger pursed his lips, "About what you'd expect for a painting lying about for over a hundred years in an uncontrolled atmosphere. As you know the remarkable thing about pigments is their stability. In addition, the varnish provided a protective coat."

"What further work were you planning?" asked Flowers.

"A final revarnishing and a new frame. The canvas is in fair shape and we won't touch that."

"What about the source of the other four?" asked Michael.

"The dealer in Arles refuses to divulge this, but Monsieur Perot has been known to be reliable and if he has a secret source he is entitled to its confidentiality," replied Sir Roger.

Michael winced, "Are you saying that you don't believe we can look for a forger by watching the dealer?"

"No reputable dealer would have anything to do with hanky panky, particularly five of the same. I personally know the man and consider him above reproach."

"May I ask a slightly embarrassing question, Sir Roger?" asked Michael.

"You won't have to, because I believe I know what it is. Part of the reason the National Gallery bought the painting was on my recommendation, and this was based on my acquaintance with the dealer. That puts me on the spot, Inspector."

"A bit awkward, Sir."

"Just between you, me, and London Bridge my position is in jeopardy. My contract is up for renewal in the spring."

"Does the dealer in Arles realize this?" asked Flowers.

"Undoubtedly," replied Sir Roger.

"I've posed that question to him and he sweats profusely. My guess is that if push comes to shove and he recognized my position, he'd come to my aid," answered Sir Roger.

"Is he protecting someone?" asked Flowers.

"Absolutely!" affirmed Sir Roger.

Landon scratched his head, "Can't we investigate at Monsieur Perot's end?"

"If necessary, of course. But we may be dealing with paintings that passed through enemy hands during the war — or even the Russians. Documentation is often impossible. You see, Monsieur Perot was on the Allied Commission after the war restoring confiscated paintings to their rightful owners. He is quite knowledgeable about that."

"How much are these paintings expected to fetch?" asked Flowers.

"They're likely to bring more than five million pounds each, if no doubt arises about their genuineness."

Flowers smiled, "Everyone has his price — your friend in Arles may have discovered what his is. It's surprising what a university professor would do for a few million quid."

"Monsieur Perot has a private collection of art worth more than ten million," underscored Sir Roger.

Flowers noticed that Michael was fidgeting to light up but didn't have the audacity to ask permission. To keep Michael distracted Robert tried another tack, "You have indicated that an investigation in Arles is not desirable — which is unfortunate since I've not been there. How can we examine the other paintings?"

"They're here in London — at Sotheby's. No other museum will bid on them until all doubt is removed. That shows how honest the dealer is. He is perfectly aware that eyebrows would be raised with so many Monets appearing out of the air. Everyone is waiting for our affirmation."

"So," said Flowers, "he could have held the other four in reserve for a few years."

"Precisely," emphasized Sir Roger. "He feels that Monets are Monets and, if genuine, should be put on the market."

"Are the other four carbon copies of this one?" asked Flowers.

"Not quite. One is painted earlier in the day, two in early morning showing hoarfrost, and one is painted at twilight. It was common for Monet to do the same scene in different light," explained Sir Roger.

"Will Sotheby's make the others available for inspection?"

Sir Roger nodded.

"Excuse me, Sir Roger," asked Michael, "What were you expecting Scotland Yard to do?"

Sir Roger grinned, "That's right, I haven't indicated our precise wishes."

Flowers grinned back, "A directorial oversight, Sir."

Sir Roger hesitated, "Is it possible to devise an entirely new technique to authenticate these paintings?"

Robert coughed and Michael turned to the redheaded professor for a response,

"Before or after tea, Sir?"

Sir Roger could respond to a leg-pull, "We are in a hurry. How about instead of tea?"

"Did you have some approach in mind, Sir?" asked Michael.

The director pointed to a specific area on the painting and gave Flowers and Landon a soupcon of a hint through his magnifying glass. Flowers examined the painting and gave a grunt to indicate concurrence.

"You raise an interesting point. I must have a few days to think about it. And before we go, Sir Roger, may I have the address of Monsieur Perot in Arles?" he asked.

The director nodded.

"One other thing — this may sound bizarre — but do you also have an address for that famous forger?"

"Sam Yates? I'm ashamed to admit that I do. We've occasionally consulted him."

"Could he paint a Monet?" asked Flowers.

"You'll have to ask him. I'm only a gallery director. He's a genius!"

Flowers and the pipe-smoking Landon parted company outside the gallery as they headed their separate ways. On the walk back to King's College Flowers caught a few minutes of a jazz saxophone at Covent Garden and this reminded him that his own group had a gig that evening. Returning to his lab, he considered the novel suggestion of Sir Roger and decided to sleep on it.

Robert Flowers arrived at King's College at 9 AM the next day, an hour early for a university professor. He rang Monsieur Perot in Arles and presented an outline of a plan to authenticate the five paintings. He needed some seed money that neither the National Gallery nor Scotland Yard would be able to provide. The dealer was sufficiently intrigued to wire Flowers the money the following day.

Robert immediately called Sam Yates and arranged to meet him for lunch at Simpsons-in-the-Strand, a bit posh for both Flowers and Yates, but why not? At least twenty-five million pounds was riding on the success of the plan.

Flowers had no idea what Yates would look like, but he scarcely expected to find an Einsteinian mass of white hair on top of a five foot five rotund, smiling, red-faced Irishman. Forgers must come in all sizes thought Robert. They shook hands and were soon seated at a corner table.

"Would you believe I haven't been here since just after the war?" said Yates.

"As best as I can remember, it's not as good," said Flowers.

"You don't appear that old. How come you have no gray hair?"

"Flowers smiled, "I have a good hairdresser."

After they both ordered roast beef, Flowers began to pump Yates,

"Are you doing much painting?"

"Mostly teaching."

"I saw a Monet recently that the National Gallery just acquired."

Yates smiled, "Is Sir Roger still worried about it?"

"Should he be worried?" asked Robert.

"Sotheby's has four more."

"How did you know?" asked Flowers.

"You can't keep that a secret. I have a friend at Sotheby's, the Director."

Flowers was amused and jumped into the business at hand, "How much to do a Monet?"

"I'm out of the business. What size and style?"

"24 by 36 — landscape, mostly trees."

The pâté arrived. "What's it all about?" asked Yates.

"I'm working with Scotland Yard — we're trying to develop a new scheme to help authentication," replied Flowers, "It's all above board."

"They all say that. £4000 if it's legal, 8000 otherwise."

"How about 3000 and use less paint?"

"O.K. — If I'm going to start eating in fancy places like this, I'll need the money."

"Here's the gig. Take a look at the four paintings at Sotheby's and paint me a fifth. You must fool the experts, after which I will advise you where you went wrong. You can then improve your effort and I'll try to detect the forgery again. The two of us will keep at it until one of us gives up. It's you against me."

"You'll use your lasers to decide?" asked Yates.

"Something like that. If you finally beat me, you get the other £1000."

"And if I don't beat you?"

"You'll have saved Sir Rogers job," replied Flowers.

"Philanthropy is not one of my strong suits, but it sounds like a good challenge."

During the entrée Yates revealed details of his early career in producing 'honest' forgeries, having always included an obvious mistake that any person with two eyes could see.

"Such as?"

"Misspell the signature, put a rude word somewhere on the canvas, that sort of thing."

"Why do you do it?"

"I like the challenge."

"Can you do a Monet that can pass muster?"

"Definitely!" Sam threw his fork onto the empty plate, "And I was right — the food has gone downhill."

"How long will it take?"

"Two months if you pay in advance, two weeks if you don't."

"Two weeks, then. Will you have some dessert?"

"Fresh strawberries and cream," requested Yates.

"They're not in season."

Yates smiled, "I don't mind waiting."

"Is there anything you'll need?" asked Flowers.

"Absolutely! I need a letter from you directing me to make a copy — we won't call it a forgery. If anything backfires, I want my ass covered. I'll need a colored photo of each of the Sotheby's paintings and I'll need money to buy an old canvas."

Flowers peeled off £50 and assured Yates he'd have the other items.

Yates ordered a Napoleon for dessert but Flowers skipped the sweets trolley and had a Cointreau. "One other thing. This is between you and me — I'll dispose of your finished canvas."

Yates grinned, "I hope you're as good as you imply with those lasers."

"I'll help as much as I can — but if I'm right there's some information in each painting that can't be duplicated, even by you."

"Will you tell me what it is?"

"Yes, but not how to do it."

"I'm never too old to learn," said Yates.

"And I'm never too old to teach."

Three days later Professor Flowers flew to Marseilles and boarded the train to Arles. His Michelin guide told him that the city was rich in vestiges of the Roman occupation, such as the Coliseum and the old Roman viaduct nearby. The guide warned that the city was difficult to navigate for its twisting ancient network of roads, although this enhanced its charm. On arrival at the station he was met and chauffeured to an impressive chateau where the Perots greeted him. Dignified, well into their sixties, and with a goodly staff of servants that formalized the atmosphere, the couple nonetheless made Flowers feel at ease.

A tour of the impressive residence further confirmed Sir Roger's assessment of the art dealer's status in society. After an elegant dinner at which Flowers apologized for not being appropriately attired in black tie, he and Perot withdrew to the study for brandies. By this time both were sufficiently at ease to discuss the Monets with some candor. Perot shook his head and pursed his lips,

"I'm afraid the Monet story does not paint a pretty picture, if I may use a colloquialism."

"Please be frank," urged Robert.

"Not to embarrass Sir Roger, I have deliberately kept some of the story from him. It goes back to the war when the Nazis were confiscating art for the glory of the Third Reich. Hitler and Goering were intent on building up their collections for the

Goering estate at Karinhall and the National Museum in Hitler's hometown of Linz. Paintings taken from Jews and not desired by Hitler or Goering were sold at auction to raise money for the war. Since Hitler thought the Impressionists degenerate, the Monets were put on the block. Unfortunately, one of our less honorable French officials bid successfully for the six."

"Six?" queried Flowers.

"Yes. I still have one. When this particular unscrupulous traitor died in 1950, his wife took possession of the estate and when she passed away a few years ago, the children asked me to dispose of them. The original auction bill of sale in 1942 proved that the paintings were bought legally, even though the price was ridiculously low. None of the Nazis wanted to be seen bidding for degenerate art. The original Jewish owners could not be traced, and even if they could, the courts would have to recognize the sale as legal since the original owners were paid for which they signed receipts. The present owners hope that their names will be kept out of the press, and I've agreed, since they plan to donate the entire proceeds to some worthy Jewish cause."

"Thank you for clarifying that," said Flowers.

"It is best to keep this story quiet since it can embarrass Sir Roger and the National Gallery. Nazi war crimes still have a sensational interest."

"Then you are serving as agent for the family?"

Perot interrupted, "I'm taking no fee — it all goes to charity."

"Have the paintings been touched in any way — cleaned, restretched, reframed — that sort of thing?"

"Not at all. The four I sent to Sotheby's are in their original frames with original varnish." Said Perot.

"That will help. Are you prepared to provide me with a letter to Sotheby's delegating me as your London representative? You can assure them that this will not alter their role as auctioneers."

"Agreed," said Perot.

"Good," said Flowers, "then all I need are some of the local tree leaves, preferably like those in the paintings."

"Come, we'll go for a walk. You may pick up as many as you wish. We have no shortage in France," assured Perot.

The stroll through the woods during the early November dusk was pleasant and the two spoke of the war, even though they were too young to fight. Perot raised an awkward possibility, "Suppose the Yates copy is detected by Sotheby's, won't that prompt them to refuse to handle the ones now in their possession?"

"Have you met Yates?" asked Flowers.

Perot shook his head, "I only know of him by reputation."

"He thinks he can fool Sotheby's and I believe him. Besides, they wouldn't doubt another painting coming from you," said Flowers.

"I hope you're right," replied Perot.

"If worse comes to worse, Scotland Yard and I will vouch for you."

"Sotheby's still won't be happy."

"Would you prefer to think it over?" asked Flowers.

"In for a penny, in for £4000. Let's go for broke — I'm fascinated with your idea."

When Sam Yates arrived with his completed painting two weeks later, Flowers was in his lab with his graduate student. Robert described the laboratory research work to Sam,

"We're examining the absorption of microwaves by superheated water. If a clean glass beaker containing water covered with a thin layer of silicone oil is placed in a microwave oven, it is possible to heat the water well above its boiling point before it violently explodes into steam. Water boils at 100 degrees only when it has a free surface to allow the water to escape."

Just as Flowers completed his explanation, the water exploded.

"Holy smokes," said Sam, "Are you working for the Army?"

Flowers led Yates to his office where he rang Sir Roger to indicate they were on their way over,

"You don't mind walking, do you?" asked Robert.

"I always take a taxi when I'm on an important job."

Ten minutes later Sir Roger was examining the two-week old Monet and shaking his head.

"No good?" asked Sam.

"I don't know how you do it. I can't find anything wrong," revealed Sir Roger who compared the Yates work with their own Monet.

Flowers removed a small instrument he had brought along, "I'll bet I can find a difference." After dark-adapting the room, he shone the battery-operated laser onto the Monet and then onto the Yates effort. With a strong magnifying glass Flowers examined the two paintings closely, "Just as I thought."

"Thought what?" asked Sam.

"Tiny little tea leaves, to quote a phrase."

After a few minutes instruction Sir Roger and Sam knew what to look for.

"Leaves fluoresce under laser irradiation — they emit light of their own when energized by the higher energy laser. There are little specks of light in the Monet but not in yours," indicated Flowers.

Sam objected, "Hold on, professor. I must be missing the point. Why are the two paintings different?"

Flowers smiled, "There are shreds of real leaves in Monet's painting but not in yours."

"I must be going bananas. Don't tell me Monet didn't use paint like me. I went to a lot of trouble to grind those pigments," protested Sam.

Removing a few leaves that he had picked up in France, Flowers demonstrated how they fluoresced. Suddenly it dawned on both Yates and Sir Roger. Monet had painted his scenes out of doors in the autumn. There were tiny particles of leaves in the air and they fell on the canvas and adhered to the oils. Yates had executed his painting indoors.

"Prof," ejaculated Sam, "You ain't just a pretty face."

"Professors are not always as dumb as they sound," quipped Robert.

"Do I get another crack at this?"

Flowers handed Sam a few leaves, "These are straight from France. Try pulverizing them and see if you can get about the same distribution as on the Monet."

Sam carefully examined the fluorescence pattern under the laser,

"This is easy."

"Very enlightening," said Sir Roger.

"Do I win if I get this right?" asked Sam.

Flowers smiled, "This is only the beginning."

Yates re-wrapped his painting and walked out whistling, "See you next week."

Sir Roger appeared disturbed, "Should you be giving your trade secrets to a forger?"

"He's given it up, and I believe him."

"He might be down on his luck and be tempted."

"I still have a few trump cards."

"I hope so — we're in your hands."

Professor Flowers was not totally surprised when Sam Yates took more than a month to reappear with his painting. Not wishing to spoil his work, he was trying to add small amounts of shredded leaf to a specially prepared six-inch sample of painting. With each try he returned and compared it under the laser to the Monet in the National Gallery workshop. He was clearly showing signs of frustration. The particles at first were too big, then were applied too liberally. Flowers was patient with Sam even though it was a bit of a fuss to accompany him to the National Gallery with each effort — Sir Roger would not permit Sam unaccompanied access to their workshop and this necessitated Flowers cutting into his own research activities to act as watchdog.

One morning a bedraggled and embarrassed Sam appeared at King's College. He sat in Flowers' office, unshaven and unkempt.

"You may have me beaten, Prof."

"You're always free to terminate the challenge."

"I know, but if I had that little determination, do you think I'd be the best in the business?"

"I wish you were one of my graduate students. They don't realize that persistence is one of the most important qualities in any research effort."

"Can you give me some help on how to get the leaves properly shredded?"

Flowers shook his head and reminded Sam of their agreement. He was not going to show him how it could be done.

"Yes, yes," said Sam, "But you didn't say you'd let me suffer."

"Suffering is good for the soul."

"That's good advice for the young, but old codgers like me have been through it all."

"Why don't you quit and blame it on the laser? You were unbeatable before it came along."

"I've licked x-rays, chemical analysis, and carbon dating — I'll beat the laser."

"Good for you. I'll look forward to your next try but we have to establish some termination date."

"When they bury me, you can say you're the winner."

"If you stay away from rich food, you're good for another thirty years."

"Not the way you're treating me. Never mind — a deal is a deal. I've one more request — may I have a few more of those French leaves? I've run out."

Flowers handed Sam four more and wished him well. Yates had grown noticeably thinner since their meal at Simpsons and Flowers had considered treating him again, but decided that fraternization was ill advised between contenders. This was a clash of giants and Flowers was enjoying it.

The Peoria Jazz quartet consisted of a piano, bass, electric guitar and trumpet. This group ordinarily served society as barrister, civil servant, gynecologist, and university professor. They rarely got paid except in expenses and applause, but they had achieved some recognition amongst their peers. Their free lunchtime gig from 12:30 to 2:00 in the foyer of the Royal Festival Hall consisted of 1920 to 1940 American jazz — light and not too noisy. Their last set finished with Rose Room, Basin Street Blues, three Gershwin tunes and When the Saints Come Marching In. They were well received by the predominantly mature audience who worked in the neighborhood and would have a snack and drink while enjoying the entertainment. The moniker Peoria had no significance other than the group leader liked the American sounding name.

With the final round of applause Flowers packed his trumpet, bid his fellow musicians adieu, and walked back to King's College across Waterloo Bridge. No sooner had he entered his office than the phone rang. A jubilant Sam was on the line, "You professors sure take a long lunch hour."

"I skipped lunch today. I don't eat before I play."

"I think I finally licked it. Are you available?" asked Sam.

"I'll meet you at the National Gallery at 3:30. Is that alright?"

"See you then."

Flowers called the gallery and ascertained that Sir Roger would be available, albeit he always reminded Professor Flowers that he felt uncomfortable with Sam inside the inner sanctum of the museum.

"I could arrange to have him handcuffed, if you like." said Flowers.

"I still wouldn't trust him."

Robert met Sam in the lobby and the two were greeted by Sir Roger who showed them into the workroom where Sam unwrapped his latest. Both Flowers and the director agreed that under laser light the two paintings were indistinguishable in respect to the leaf fluorescence.

"It was that experiment you did with the microwave that gave me the idea. I was down to the last leaf when I borrowed a friend's oven and slowly evaporated the moisture in the leaf. This left the leaf dry and brittle and I ground it in a mortar and pestle.

I removed the varnish and sprinkled some leaf onto the painting through the pepper mill. What do you think?"

Flowers examined the painting more closely under a microscope and decided Sam had achieved success.

"Very good, Sam, I'm proud of your persistence." Flowers patted him on the back and then drew out his other laser, "We're now going to examine something Sir Roger suggested." Shining the laser onto the section of the Monet first noted by Sir Roger, Flowers pointed to a short length of brownish hair. "That, gentlemen, is probably a small length of Monet's beard."

Sam and Sir Roger took turns checking it through the magnifying glass. Sam viewed Flowers with disbelief, "Don't tell me you've brought back a sample of Monet's beard?"

"I trust Monet will rest in peace and I won't have to resort to body snatching. However, this higher energy laser is better for hair fluorescence," advised Flowers.

"Holy smokes," exclaimed Sam, "I give up. What am I supposed to do now, find a French artist with a beard like Monet?"

"You could try," chided Flowers.

"This isn't cricket. You've obviously been putting me on," complained Sam.

"Right now," said Flowers, "we have five Monet paintings with small bits of hair which I doubt we'll see in your work. Even if a few of your gray hairs have fallen in, they've lost their pigment and won't fluoresce the same. If Monet had been older when he did these paintings, you might have been lucky."

"I quit! For all I know you've got some more aces up your sleeve."

"Think it over," urged Flowers.

Sam could not be placated. He took a quick look at his own painting with the higher energy laser, shook his head, re-wrapped his work and left without saying goodbye.

Sir Roger was pleased but apprehensive over this turn of events. He told Flowers he'd be much happier to know that there was something that Yates could not duplicate, "I'm worried. Yates is driven by a big ego."

A week later a confident Sam Yates returned to King's College to demonstrate the addition of small fibres of brownish hair under the laser. Flowers examined the effort and agreed that Sam had succeeded.

"How did you do it?"

Sam related that he had taken out several books on Monet looking for any reference to his beard colour around 1875. Photographs of Monet were all in black and white although he did find a coloured portrait reproduced in one of the books. Sam collected locks of hair amongst his friends and finally found the exact match. Right now he told Flowers he had no friends left.

"Let's have lunch at Simpsons to celebrate," suggested Flowers.

"We've already eaten there. Couldn't we try the Savoy?"

"You're developing expensive tastes."

"I've always been that way when someone else is paying."

Flowers agreed to the change in locale and the two strolled down the Strand to the hotel. Sam forgot that the short street down to the hotel is the only one in London on which drivers stay on the right. It was adapted as a convenience to taxi drivers so they could open the passenger door without leaving their seat. Sam was almost killed looking the wrong way. To calm his nerves he ordered Beluga caviar, lobster, and their finest champagne. Flowers sensed that Sam was deliberately trying to be provocative.

"It's nice to know I've beaten the experts again — not that I'm interested in returning to the old game. I wouldn't have anything to do with forgery." Sam ate with relish and savoured the champagne, "I'd love to become an expert in wines. Of course, it takes a little capital to develop an educated palate."

"I hope you don't throw away your £1000 on drink."

Sam smiled, "With what I know now I could turn out a new Monet in about 10 days complete with leaves and hair. It takes a long time to learn how, but once you know the ropes it's a cinch."

"It's no different in scientific research," underscored Flowers.

"In fact, I could train a young art student to churn out a Monet in no time — complete with leaves and hair."

Flowers eyed Sam and thought the champagne might be having some effect. Why did he repeat himself?

"Let's see," said Sam, "What looks good for dessert?"

"You've been most generous with my business account. Shall I ask how much the waitress charges?"

"I'm a little old for that, but thanks for thinking of me."

Sam settled for Peach Melba with some green Chartreuse. When it arrived Sam reminded prof that Chartreuse is made in southern France, Monet country. "Perhaps Monet spilled some on his painting."

"For years Monet was too poor to afford such a luxury."

"I hope I'm never that poor again," said Sam.

"If you leave the painting with me I'll have a cheque for £1000 in the post tomorrow."

"It's been good fun working with you, even though I wasted so much time over those leaves. Still, it was a rewarding experience."

The two returned to King's College, and Sam bid goodbye after placing a note in Flowers hand.

"What's this?" asked Flowers.

"It's a surprise. You've taught me a lot — I want to teach you something."

"Thank you. If I don't see you again, enjoy yourself."

"I intend to, prof."

As Sam disappeared from view Flowers unfolded the note and began to read it. After scanning the first sentence, the phone rang.

"Professor Flowers, this is Roger van Zaal. There's a Board of Directors meeting next week and they'll want a briefing on the status of the Monet. They are, of course, unaware of your experiment and the individual involved, heaven forbid. How much longer is it likely to be?"

"He's just left and I'll be sending him a final cheque. I hope you've been satisfied that we've managed to find two bits of confirming evidence to support the authentication. I needed the challenge of comparison to illuminate the small differences with the leaves and the hair."

"Yes, I agree that your laser technique has given us more confidence and you've fulfilled the original request. I'm still not happy that you have such a perfect forgery. Will you destroy it?"

"I was planning to hang it in my office. Something will be done to ensure that it won't be mistaken for a real Monet," assured Flowers before ringing off.

Sam's handwriting was atrocious although the message became obvious:

Dear Professor Flowers,

I derived great pleasure from the challenge you presented me, particularly in the satisfaction of beating the experts again. While I've sworn off performing forgeries, I believe my services as a consultant have not been adequately rewarded. Hence I'm suggesting that I be used as an independent authority in the future disposition of Monets containing leaves and hair. My fee will be modest — only one percent of the auction price.

With great respect I am your humble servant,

S/ Sam Yates

Flowers reddened at this veiled threat of blackmail. One percent of twenty-five million was two hundred and fifty thousand pounds! It was clear that Sam could do considerable damage by informing Sotheby's or the Ministry of the possibility of producing a forgery. In fact, that one already existed was sufficient. After all, Sam even had a letter requesting him to perform this service.

Red hair is reputed to be a sign of a violent temper. For thirty-five years Professor Flowers had managed to remain placid dealing with students but at this point in his career he discovered this latent rage. He tried phoning Sam but he had obviously not arrived home yet. He slammed the receiver down and began pacing

Q corridor. Damn, damn, damn! He thought. Flowers left the College by the Embankment exit and walked along the Thames to cool down and think what to do next. He cursed his naiveté and wondered how he would break this disastrous news to Sir Roger. After walking to Blackfriars Bridge and back, he had spent some of his fury and returned to his office where he managed to reach Sam on the phone,

"Thank you for that lovely note but I believe your math is a bit in error. You probably meant one hundredth of one percent — £2500 not £250,000."

"I may be stupid at times, but never with my math," said the devilish voice.

"This is pure blackmail and we'll have nothing to do with it."

Flowers slammed the receiver down and took out his trumpet. Locking his door he played the Flight of the Bumble Bee, an exercise that caused his veins to bulge and relieve his tensions.

For several days Robert became incommunicado as he alternated between walking and trumpet exercises. At the end of the second day the germ of an idea emerged and he returned to re-examine the Monet at the National Gallery. Taking a series of holograms — photographs under pulsed laser light — he repeated the process on the Yates painting and approached the head of the Image Processing group with his problem. The group under Professor Budge did computer analysis of photographs to bring out details not visible with the naked eye.

With the clue given them by Flowers the group achieved quick results. Ringing up Sotheby's Robert obtained permission to make holograms of their four Monets after first requesting them to clean the paintings in order to remove surface fingerprints.

A week of detailed analysis produced the information he needed to verify his theory. Another set of holograms were required to concentrate on selected areas of the paintings and he repeated this measurement on three well known Monets hanging in the National Gallery before he had convincing proof.

He did not divulge the nature of the crisis to Sir Roger, nor did he accept any phone calls from Sam dunning him for an answer to his request. He had to be certain of his ground before revealing the details to Sir Roger and telling Sam to shove it.

Flowers was beginning to savor the sweet prospect of revenge and he phoned Sam at 6 AM on Sunday to give him a doubly rude awakening.

"Professor! I've been trying to reach you all week. I hope you're not mad at me."

"Why should I be mad at you? I called you on a Sunday to remind you to go to church and pray. I'm as happy as a meadowlark, now that I've beaten you again. This time I'd say you've lost it all, including your thousand."

A good ten seconds of silence indicated that Sam was still half asleep and trying to make a rapid assessment of this turn of affairs.

"You're trying the old bluff, are you?"

"If you take that attitude we'll terminate our discussion and I'll bring Scotland Yard in. Any further attempt at blackmail, no matter how veiled, will find you in the dock. I assure you, it will be better if you cooperated."

"I still think you're bluffing, but I'll give you the benefit of the doubt, Professor."

"Whose doubt?" asked Flowers. "If you come into my office tomorrow at 3 PM you'll see what I'm talking about."

"I'm a little nervous about things like tape recorders."

"You won't have to say a word — just look and listen."

"You don't have a new laser for reading people's minds, do you?"

"I don't need a laser to read your mind."

"I see what you mean," said Sam.

"Don't forget to go to church."

Flowers gently put the receiver down and said a little prayer for the Divine Guidance he had received during the week.

Sir Roger and Sam arrived at the same time the next day and were shown into Professor Flowers office by the secretary. Tea was served and Robert got down to business,

"Frankly, I'm glad Sam threw down the gauntlet for something more convincing than the leaf and hair fluorescence. You may be aware that lasers have proven valuable in fingerprint detection, the oil residue from the fingertips shows up nicely under laser fluorescence and avoids the tedious and less efficient method of using a fine powder and a brush. The powder method is less desirable than the non-contact laser technique.

Sam's expression showed signs of extreme tension. He was not likely to call Flower's bluff again and paid close attention to the professor.

"By careful laser scans and photography of seven Monets and the one Sam produced, I've now established the thumb print of one Claude Monet and a complete set of Sam Yates' prints, which I compared to the set on file at Scotland Yard."

Sam coughed, "Must we go into that, Sir? It is past history, is it not?"

"In a way these paintings represent past history," quipped Flowers.

"I take your point, Professor."

"The Monet thumb print has gone into our confidential file but I'm willing to give Sam another chance to work on that point — on his own."

Sam considered this impossible challenge for a few seconds and headed for the door.

"Hold it, Sam. Can't I buy you lunch? There's a terrific McDonalds on the Strand — ten queues, very little waiting."

Sam eyed Flowers momentarily and stormed out the office.

Sir Roger looked at the image analysis reconstruction of the Monet thumb print, "Intriguing! This could prove to be a decisive tool in authentication."

"Actually, I thought I was beaten when I discovered that two of the Monet paintings showed no prints."

"Which two were they?" asked Sir Roger.

"The early morning paintings showing the hoarfrost. I spent a sleepless night on that."

Sir Roger shook his head, "No doubt there's a simple explanation having to do with temperature and that sort of thing. I suppose the oils from the fingers do not adhere?"

"Not when the artist is wearing gloves."

Sir Roger and Flowers had a good laugh. When this spell of jollity abated, Sir Roger pointed to the Yates' Monet hanging on Robert's wall, "What are you going to do with that?"

"I had Sam forge me a letter that I've pasted to the back of the canvas."

Flowers removed the painting and turned it over for Sir Roger's inspection. A letter was affixed that read:

THIS PAINTING WAS EXECUTED BY A DISCIPLE OF MINE IN THE STYLE ATTRIBUTED TO ME. HIS NAME IS SAM YATES AND I COMMEND HIM TO THE ARTISTS OF THE WORLD.

S/CLAUDE MONET, Dec. 14, 1991

ENGLAND: 1980–1990

Richard contacted Beecham's, a pharmaceutical company near Guildford and induced them to put up some money for a graduate student to work on a metallurgical problem of interest to them. They selected the question of shampoos and the effect they had on hair. It seemed that the public wanted a shampoo that left the hair squeaky clean. A surfactant was added to the shampoo to dissolve the natural oil secreted by the scalp to coat the hair. The shampoo left the hair dry, lackluster, and unmanageable. Beecham's then sold the consumer a conditioner to replace the lost oil and restore the sheen. Question: On a microscopic scale what did the hair look like after shampooing and after conditioning? And where did these products actually go within the interstices of each hair?

These questions could open the door to a more ambitious research program — just where it might lead after the work began was a matter of serendipity and hard work. Looking through a microscope Richard was fascinated with the overlapping scaled appearance of hair, like an armadillo. A graduate student was identified who agreed to work on the project but alas! before long she changed her mind. No one else could be found to carry on the work and the project never got airborne. But one interesting unsolved problem caught Richard's fancy, the atomic and genetic mechanism responsible for natural hair color, particularly red. With billions spent on products to alter the hair's appearance one might have thought that some research would have made inroads into this area. Richard never found the opportunity to pursue this question and it dropped by the wayside together with much of Richard's hair.

When in Avon Richard worked out of his home. He received a call one day from Robert Clark, editor of a magazine *LASERS AND OPTRONICS*. Would Richard write a feature article entitled FROM THE LABORATORIES? Richard agreed and once a month he would visit a university or industrial research team and determine the gist of their activities in optoelectronics. He'd then compose a thousand words in a popularized form that would not bore those readers with limited knowledge of the subject. This was fun because it presented a challenge to Richard's scientific and literary ability, for example:

From the Laboratories
WRITE YOUR NAME ON A MOUNTAIN

"Tucked away south of Boston in a colonial house, circa 1700, is a company that produces a 'user friendly' display system that permits one to write, direct, and produce an entire animated laser light show for educational, display, or presentational programs. In less than a few hours the user can learn the essential of the new Lasergraph's controls and begin to devise a visual display that moves, responds to audio or other cues, activates kaleidoscopic projections and other eye-appealing artistry, and even delivers coherent messages. At a price tag of around $20K ($2500 for a single event rental) it enables anyone to get into 'show biz', all in the name of promoting your firm's specialties.

Brainchild and founding father of Roctronics Lighting, Lasergraphics, and Special Effects, Inc. is Dr. Richard Iacobucci, a man who has put in his time backstage and who has become expert in lighting displays, both conventional and laser. With a B.S. in electrical engineering, an M.S. in computers, and a PhD in law he appears qualified to take on all challenges. The company earns its bread and butter by selling controlled incandescent lighting displays and special effects, in addition to which Dr. Iacobucci will design optical pyrotechnics to meet the specific needs of a theater, a nightclub, an advertising display, a restaurant, a conference center, trade show, etc., the venture into laser manipulation was a natural outgrowth of his talent.

The Lasergraph™ itself is a portable lightweight control panel, about 18" × 36" × 4", with 100 manual control buttons, joy sticks, and switches that employs a basic language of more than 4000 solid state memory-stored figures and words that can be tailored for the special needs of the customer. Such images as stars, flowers, butterflies, animated cartoons, logos, faces, words, letters, etc. comprise the basic image 'language'. The controls such as size, roll, pitch, yaw, zoom, audio response, etc. permit one to make the basic language images fly, move, reverse, grow, wane, and sequence themselves in any desired way — you name it. Cueing signals can activate ancillary displays of any kind such as smoke effects, bubblers, strobes, colored lights, etc. Finally, a memory bank accepts user-created instructions so that once you've staged and rehearsed your show, you can sit back and enjoy the fun of watching your audience "ooh!" and "aah!" at your talent as the laser beam demonstrates your wizardry as an artist and storyteller. When you display the final credits of WRITTEN, DIRECTED, AND PRODUCED BY Joe 'Spielberg' Zooch, you can modestly nod, smile, and say, "I want to thank LASEROPTOTRONICS, Inc. for supporting me when others pronounced me redundant, etc., etc., etc."

Iacobucci seems to have created a laser control technology that has responded to three desirable constraints; make it simple to operate, make it inexpensive, and make it easily portable. Laser light shows, like the Laserium reported earlier involve considerable expense and operating talent and are more or less permanently bolted down.

With a built-in audio response to the amplitude of sound waves one can animate figures with the tempo of music. A basic heart-shaped image decorated with the waveforms of a musical stimulus through the audio size-modulation feature can be created. Image duplication and rotation is accomplished by motor driven prisms mounted in the optical head of the Lasergraph system. It can also produce a circle of duplicates, where the center image of a butterfly remains stationary while the clones rotate about it. By sequencing the butterfly closed-wing image with that of the partially open and fully open images, etc. the wings appear to flap, and the butterfly to soar about in circles. An obvious extension of this is to have Roctronics load your company logo into the image memory endowing it with wings or rockets so that it may fly through your presentation and, in lieu of a pointer, momentarily rest on something you wish to emphasize.

Another feature permits one to compose messages on the spot by instructing the memory to combine alphabet characters. Special user artwork can be duplicated by Roctronics and stored into the 4096 images' memory.

While a small image can be created with a 5 mw red HeNe, the Lasergraph optics can as easily handle 10 watts of a blue green argon laser for large mountainside displays. The four-color displays of the argon-krypton laser employed by the Laserium is not part of a basic Lasergraph but can be accomplished by employing several Lasergraphs working in parallel.

Another application is to use two-track audiotape, one of which provides music and narration, while the other provides cues to synchronize the visual Lasergraph display. But, as frequently happens in such versatile instruments, Dr. Iacobucci predicts users will devise additional unique programs and techniques beyond those envisioned by the inventor."

Alas! Lasers and Optronics ran into hard times and changed its format after two years. Richard was again up the creek without a job. However Joe Yaver came to the rescue and Richard was engaged for 4 years to write a similar monthly article, to be called THE RESEARCH SCENE, for the SPIE newspaper.

The soft-spoken and competent Professor Bill Sherman at King's College was both an optics and a high-pressure expert. He had developed a special cell that permitted optical examination of the sample as it was squeezed hydraulically between two vertical pistons by providing two small horizontal access ports that were filled with an optically transparent material. Richard had found some US Army research funds to enable Sherman to examine the compressibility of TiB and TiB_2, both hard armor materials. The technology required an energy dispersive x-ray unit, which Richard located in the Physics Department at the University of Munich, courtesy of Professor Peisl. As one increased the pressure on the sample the atoms were pushed closer to each other and this could be determined from the change in the scattering angle of the x-rays. The lower the compressibility the more pressure was required to push the atoms together, hence the better its resilience as an armor.

Richard hoped to kill three birds in Munich. In addition to carrying out the US Army research program he planned to measure the thermodynamic properties of vanadium, particularly the frequencies of atomic vibrations. As a side issue Richard had been developing a story about the German scientists that worked on the atomic bomb during WWII. Some of the story took place in Munich.

Most people identify Munich with beer halls, like the Hofbrauhaus, with Bavarians in lederhosen, with Hitler's beer hall putsch, with the Octoberfest, and with 'Crazy' King Ludwig's Disneyland Castle south of the city. Few people, including Müncheners, realize that their huge Englischer Gartens, one of the largest parks in Europe, was conceived and built by America's second physicist Benjamin Thompson (Franklin was the first). Born in 1753 in Woburn, Massachusetts, he grew up a poor boy with a cruel stepfather but was fortunate enough to be apprenticed to a doctor who saw to it that Rumford was given a modest education. As a handsome redheaded lad of 19 he married a wealthy widow ten years his senior and soon developed delusions of grandeur with his newfound wealth. He remained a Loyalist and spied for the British until Washington threw the Redcoats out of Boston whereupon Benjamin sailed for England. Convincingly boasting of his accomplishments Lord Germain appointed him as a consultant for American affairs but he left for Bavaria after mismanaging his duties. His braggadocio had become insufferable and Germain was glad to see him leave.

The Bavarian King was impressed with Thompson who boasted of his American accomplishments and made him his military advisor. Thompson began his experiments on the mechanical equivalent of heat by observing the temperature rise when boring gun barrels. He was appointed General in charge of the Bavarian army and saved the city from invasion by the French and Austrians. The King rewarded him by elevating this commoner to the position of Count of the Holy Roman Empire (Count Rumford). Thompson then conceived the idea of turning the King's hunting preserve into a park to provide gardening opportunities for all the beggars he had taken off the streets of the city. However he alienated so many of the King's counselors that he was forced to leave Bavaria.

The University of Munich was located in the Schwabing section of the city, a district with a Greenwich Village atmosphere. Bill Sherman and Richard got down to work and completed the preliminary results on the armor materials. Amongst the hard materials in nature diamond ranked the most difficult to compress with TiB_2 a close second. TiB was about 30% higher, still a good armor candidate.

Richard gave a talk to the physics department about Count Rumford. The faculty were all aware of Rumfordstrasse and had heard of Rumfordsuppe but didn't know the significance of the special name for this soup. Thinking like an experimental physicist when he rid the streets of Munich of thousands of beggars Rumford had tried a number of concoctions to determine the cheapest and most satisfying meal to feed these people while they were put to work making uniforms for the soldiers. Rumfordsuppe consisted of beans and other pulses, flavored with vinegar and served with croutons to stimulate better mastication.

If Rumford hadn't alienated so many people during his infamous career he'd be better known today. In his lifetime he endowed a chair in physics at Harvard University, arranged for the annual awards of the Rumford Medal, built the Royal Institution in London, and even saw to it that Harvard would tend his grave near Paris in perpetuity. Richard called up the occupant of the Rumford chair at Harvard and found out the man knew virtually nothing about his benefactor.

After Sherman left Munich Richard continued the work on vanadium. This involved measuring the intensities of the diffraction lines as a function of temperature from the lowest temperatures available to him to room temperature. The work was later combined with neutron diffraction measurements on vanadium by Clive Wilkinson of the physics department at King's College and then published. Differences were found between the spectrum of vibrational frequencies measured with neutrons and those measured with x-rays. Alas! No one appeared interested in this negative result and the measurements have not been repeated. Richard concluded he was either jinxed or had become asocial.

In recounting the sequence of Richard's interests in solid state physics he began by trying to measure the electron distributions in solids like iron. His limited success led to measuring the electron momentum distributions by Compton scattering. Here the revelations were considerable. Once these distributions were determined for a solid whose atoms were 'at rest' the question arose as to how the thermodynamic properties could be determined from the electron distributions. The electrons not only bind the atoms together but control the thermal vibrations as the temperature is changed. The frequencies of these collective vibrations (phonons) determine the specific heat, the phases present at any temperature, and various other properties in metallurgical processing. Unfortunately Richard had reached another impasse in his effort to find ways to measure these frequencies.

One of Einstein's important contributions to physics was to show that the energies of the phonons had to be regarded as quantized just as the energies of electrons on atoms. With this model he calculated the low temperature specific heat of a solid, until then a theoretical puzzle. But life was never that simple since the nature of the forces between atoms was so complicated. It was well known that it took more energy to push atoms together than to pull them apart an equal distance. This was called anharmonicity. As a result the quantized energy levels were no longer simply spaced a la Einstein but forced the phonons to interact with each other. In a paper he published in 1963 Richard examined a simple model of a vibrating atom to show to what extent anharmonicity 'buggered' up the simple picture employed by Einstein. Theoreticians turned a cold shoulder to anharmonicity since it complicated their calculations to the point of hopelessness. That is the rationale that drove Richard to develop an experimental technique to examine materials and look for theoretically tractable simplifications. In his work on vanadium it appeared that anharmonicity was a likely culprit to explain the disparity in the neutron and x-ray results.

As Richard realized, theoreticians didn't like complications and anharmonicity was the devil incarnate. It was the responsibility of the experimentalists to provide clues which would aid the theoreticians in making their calculations. It is understandable that theoreticians must produce papers — that is their bottom line. Hence the experimentalist must wean them away from models that are too simple to illuminate the workings of nature.

On a courtyard wall at the University of Munich Richard found a plaque honoring the 1913 Nobel Laureate Max von Laue for his discovery of x-ray diffraction by crystals. He was an outspoken critic of Nazi policies, even helping Jewish students by giving them his ration cards, yet he survived the Nazi purges. Whether this had anything to do with his standing as a Nobel Prize winner or with friends in high places was unknown but when Sam Goudsmit selected those German atomic scientists in 1945 to be detained in Cambridge England he included von Laue even though he had nothing to do with uranium work. Those days at Brookhaven in the early 1950's when Richard spoke to Goudsmit about the ALSOS mission had provided the stimulus to get at the truth. Richard was not the only one after this story but he appeared to be the only research physicist who had tried to produce a play about it.

The problem about a play was that one had to create character. Of the ten scientists who were interned at 'Farm Hall' Richard selected five: Hahn, von Laue, Heisenberg, Gerlach, and Diebner who he considered the most likely to move the story along. Having only met Heisenberg Richard had to create dialogue that made each 'voice' clear and identifiable. In addition the narrative had to move along with proper development and character exposure. He had hoped to get further insight into the five characters by having people who knew the characters read the play's first draft. Peisl arranged a meeting at Frau Gerlach's flat in Munich and invited Hahn's grandson and biographer, and one of Heisenberg's students.

Frau Gerlach was gracious enough to provide a dinner and the group imbibed quietly as they waited for the after-dinner conversation to unleash the comments. Richard was attacked on all sides since no one but von Laue was portrayed as a hero. Walter Gerlach was known to have had an abrasive temperament and as leader of the uranium project may have been the right man. People did their utmost to avoid his outbursts. Frau Gerlach supported her husband's noble character by providing Richard with some reprints illustrating Gerlach's admiration for Einstein. But when two of Gerlach's aides at the University told Richard of the time he was so enraged after a phone call that he tore the instrument from its connections and threw it through a closed window. When Richard was informed that Gerlach was a terrible experimentalist and was better employed as an administrator, when Richard learned of Gerlach's passion for roses and the piano, when Richard was informed that Gerlach considered suicide for failing the Fatherland after learning about Hiroshima, he felt he had established some basic characteristics to employ in the play.

Hahn's grandson was as disturbed with Richard as Frau Gerlach but he convinced Richard that his grandfather was not a Nazi sympathizer. Hahn had exerted himself to help the Jewess Lise Meitner escape Germany and had become overly distraught over the realization that his discovery of fission had led to the atomic bomb. No one questioned the von Laue or Diebner characterization, one good the other questionable. The enigma was Heisenberg, the brains of the program. A close friend of Heisenberg and one of the ten interned scientists was living near Lake Starnberg south of Munich. Carl Frtedrich von Weizsäcker had been a young theoretician when he worked with Heisenberg on the uranium problem. His father was State Secretary under Hitler and had to face charges at Nuremberg. His younger brother was

President of Germany in the 1980's. The stage play about the internment in England was sent to C.F. von Weizsäcker and he assented to see Richard at his home. Several hours were spent during which corrections were suggested to Richard and a general discussion held about the uranium project. Richard returned to Munich, sent a corrected manuscript to von Weizsäcker and arranged for another appointment. This time his host hit the roof saying that he had hoped Richard would have gone away and dropped the project, Goudsmit didn't know what he was talking about, etc. Why was he angry? Richard never found out but von Weizsäcker affirmed what Heisenberg had claimed in his running battle with Goudsmit — there never was a German effort to make a bomb. Richard concluded that this was technically true since the German effort never got far enough. If they had constructed a working reactor history might have been different.

Thus ended Richard's search for the truth in Munich. In London he discussed the story with R.V. Jones, the scientist who had arranged for the 'bugging' of the home where the Germans were interned, and with Sir Michael Perrin, former director at Harwell and member of the ALSOS mission. After he rewrote his play for the third time Richard convinced the BBC to do a radio version of his stage play. It was aired while Richard was in Avon and he had little feedback. Richard concluded that it would have been impossible for the Germans with their limited resources to have mounted an Oak Ridge-Hanford-Los Alamos type effort since it would have been bombed out of existence in 1944–45. Was Heisenberg deliberately dragging his feet to keep Hitler from getting the bomb? The successful sabotage of the heavy water source in Norway prevented Heisenberg's group from constructing a reactor (they thought graphite would not work). The German group never felt a sense of urgency as the Americans did — their leaders never had to make engineering decisions about building a bomb.

Richard's fetish for popularizing science and his interest in Lewis Carroll, story teller and mathematician, prompted him to try writing a small volume about physics employing characters from

'Alice in Wonderland' and disguising the names of some physicists. Macmillan UK agreed to publish it. The book was written as a stage play to be read, not performed. If it aroused interest in physics that was all Richard asked. Unfortunately Macmillan USA chose not to distribute it and the slim book never had a shot at the American market.

"THE MAGIC OF PHYSICS
Can you pull a Rabbit Out of a Black Hole?

BACKWORD
Can one learn physics falling down a rabbit hole? Even with a firm hold on your metre stick and stopwatch, your own screaming will destroy your concentration and you'll have to rely on someone else to measure your acceleration due to gravity and your terminal velocity.

Ah! — but falling up a rabbit hole and out into the wonderland of nuclear magnetic resonance, non-destructive testing, laser weapons and orthogonal wave functions — that's real physics!

When Alice, the Mad Hatter and the Caterpillar are hired by Humpty Dumpty (alias Max Wells) to pursue physics on board the research vessel PUBLISH OR PERISH the improbable thing happened — Max Wells' timely murder right between the covers of this book.

Is this just another crime? If you turn to the appendix and examine **20 Ways to Show You're a Physicist** you may recognize yourself. If so, you'll know the killer. If not, you've a lot to learn — you took a wrong turn up the rabbit hole. This book is required to bring you back by the hare.

Llorrac Siwel

ACT ONE
Scene One
(Alice and Professor Schrodenberg are walking in the Legendre Gardens. Alice picks up a white stone.)

ALICE And what does quantum mechanics tell us about this beautiful rock?

SCHRODENBERG Not very much — not very much.

ALICE Even its colour?

SCHRODENBERG Not even that.

(Alice picks up a fallen rose petal)

ALICE Surely something about this rose?

SCHRODENBERG That's more difficult than the rock.

ALICE And this tree?

SCHRODENBERG I believe it's oak.

ALICE But isn't that elementary?

SCHRODENBERG L M N tree? Funny, I thought it was oak.

ALICE You told me that wave functions had to be orthogonal — why is that?

SCHRODENBERG Now <u>that's</u> elementary — if they weren't — why you'd disappear in a cloud of smoke. Remember, your ground state must be well-behaved.

ALICE Is that why people should not get excited?

SHRODENBERG A littler perturbation is alright — but only if their ground state is still recognizable.

ALICE And all this was discovered in 1926?

SCHRODENBERG Yes.

ALICE How did people manage before then?

SCHRODENBERG The world was very confused.

ALICE But we <u>still</u> have wars."

The story line continued through trials, murder, research, etc. The appendix added a few bon mots:

"APPENDIX
20 Ways to Show You're a Physicist

1. Never answer your mail. 99 percent of those who write to you are in an inferior position. Why else would they write? However, requests for reprints of your articles should be answered immediately to ensure rapid propagation of your brilliance.

2. When reviewing a book be certain to find at least four errors in judgment or fact. Remember, as a book reviewer your superiour position has already been acknowledged — it is your task to justify that trust.
3. Never address an audience without a piece of chalk in hand. The listener's eyes are glued to this weapon waiting for it to strike.
4. Never give a talk without a mathematical derivation.
5. If anyone in your audience is not taking notes, glare at him unmercifully. The gems of your oral utterances deserve to be recorded for posterity."

Etc., etc., etc., etc.

"The key to proper behaviour is arrogance — God's way of saying he didn't make us all physicists. It is a small price to pay for the atomic bomb and television."

While in Germany Richard became interested in the story of Therese Neumann, a girl who lived in the East Bavarian town of Konnersreuth and went into religious ecstasy every Friday, displaying the stigmata on the hands, head, and feet that corresponded to the Crucifixion. In particular, was the physical evidence clear? He turned it into a detective story, the protagonist being a priest/doctor who had been assigned to explore the medical evidence. By a queer turn of fate Richard discovered that his cousin, a psychiatrist, had written several papers for a medical journal on the subject of stigmata.

But once having wandered into the field of the paranormal he became intrigued with the occasional efforts to explore the subject with scientific precision. He found a physics professor at the University of London who was intrigued with the spoon-bending craze of Yuri Geller. Richard was invited to observe his experiment.

Richard entered the laboratory in the physics department and was shown an aluminum rod hanging from a support and connected to a thermocouple and a strain gauge. A particular subject, a youth of about 20, arrived and projected his thought impulses toward the rod, never being permitted to touch it. Any rise in temperature or slight bend in the rod would have been recorded on the chart. Richard watched but nothing happened for an hour. He then made some excuse to leave.

Shortly after the tavern opened, when Richard found himself alone in the building, he heard footsteps from the second floor. He was too frightened to investigate but when it happened again he mounted the stairs with heavy tread to explore the upper floor. Nothing could be found. Several other trustees had experienced strange sounds when alone in the building and the rumor circulated that the tavern had a ghost. A tavern patron who professed expertise in these matters was permitted to investigate the second floor with Richard close behind. The woman frequently stopped to comment that several spots unmistakingly reflected some 'aura' of something or other. After an hour Richard gave up and left the woman to continue her search. No ghost has ever been seen but the rumor of one has not died.

A chemistry professor at King's College London gave a talk about his thirty year search for ghosts. He had amassed several pieces of equipment including infrared camera, sensitive recorder, thermometer, etc. He would install these at any locale where ghostly emanations were reported. His conclusions after thirty years were that ghosts did not exist or, if they did, they would not reveal themselves when he was in the house. This latter possibility can not be ruled out as mere humor.

The writings of Benjamin Franklin, particularly the aphorisms from Poor Richard's Almanac, made for humorous reading at the Tavern:

1. There are more old drunkards than old doctors.
2. Fear not death, for the sooner you die the longer you'll be immortal.
3. If your girl laughs at everything you say she probably has a fine set of teeth.
4. If a man could fulfill half his wishes he'd easily double his troubles.
5. Keep your eyes wide open before marriage, half shut afterwards.

Having become an avowed Franklinphile Richard wrote a monologue for an aging Franklin. This was performed on BBC Radio 4 and achieved some distinction with its selection as Pick of the Week. Encouraged with this success Richard began to delve into biographies and produced over a dozen monologues. Several of these have been performed at the tavern.

EPILOGUE

Watertown has had its innings. It became a center for a worldwide experimental and theoretical effort at identifying electron behavior on atoms. But by 1980 the group was forced to turn to other applied US Army problems and by 1990 the decision had been made to close the Arsenal. Developers will someday be bidding for this valuable piece of land. As the last piece of the reactor is hauled off to the dump together with the research lab, Richard will probably invite his comrades to a wake at Blanchard's Tavern.

Still alive, though, are the Sagamore Conferences and the memories of those 30 years from 1950 to 1980 when the electrons in selected materials felt the hot breath of x-rays and neutrons seeking them out, wherever they were.

HAIL, NOBEL ELECTRON

I hid from man's gaze
for an eternity. Watertown
helped bare me to the world.
The Arsenal is gone but I
charge forward with full
momentum, my head spinning.

After J.J. Thomson discovered the electron a parody to the new particle was composed to the tune 'Clementine':

In the dusty laboratory
Mid the coils and wax and twine,
There the atoms in their glory,
Ionize and recombine.
Chorus
Oh my darling, oh my darling,
Oh my darling ions mine.
You are gone and lost forever
When just once you recombine.

AVON: 1990 –

Once he retired from King's College London Richard's physics activities concentrated on optics, amusing himself by visiting laboratories and converting the scientific output into short, short stories for publication in the Lasers and Optronics magazine and later in the SPIE newspaper. But the wheels that generated ideas kept meshing in his cranium and he'd scarcely report on a piece of research without relating it to other efforts that he was aware of. And so emerged the idea for Sensored Composites, originally conceived at King's College.

Graphite fibers had become the material of the future because they had a high modulus of elasticity, i.e. it took a very large force to stretch them. When embedded in epoxy the resultant composite structure was both stiff and light weight, finding its way into golf club shafts and tennis racket frames. McDonnell Douglas fabricated an entire plane's tail section to test it for in-service reliability. Someone followed on with the idea of embedding an optical fiber into the graphite composite to measure the strains from changes in the light signal passing through the fiber.

One day the idea clicked. Why not use strong optical fibers to fabricate the composite thus providing an optical feed through to record the strains in the composite. Richard found an angel and a patent application was prepared through an attorney. The press release told the story.

Press Release Mar. 31, 1990
AIRCRAFT INSPECT THEMSELVES

Professor Richard J. Weiss of King's College London announces a novel concept in aircraft structures based on a new type of opto-mechanical material called Sensored Composites*.

With the development of optical fibers as strong as steel, a structural composite is formed by embedding a bundle of these fibers in a matrix, like epoxy, and continuously monitoring the optical signals launched at one end of the composite and detected at the other end. The fibers provide both the strength and optical path in the Sensored Composite.

When employed as the structural members of an aircraft one can continually detect the twisting and bending of the structure since the optical signal follows these changes. Furthermore, any cracks that develop break the individual fibers and prevent the optical signal from being transmitted. The periodic out-of-service x-ray and ultrasonic inspection of aircraft is eliminated since the integrity of the structure is continuously monitored from its virgin state when installed, to its long term state as the aircraft builds up flying hours.

During flight the twisting and bending of wings and fuselage is continuously recorded. When any part of an aircraft shows the first signs of abnormal behavior, that structural component can be replaced. The out-of-service testing of aircraft is time consuming, costly, and is performed on a stress-free structure. When an aircraft is fabricated from Sensored Composites the in-flight integrity is recorded under conditions of high stresses, i.e. when the entire weight of the aircraft is supported by the wings.

Another option with Sensored Composites is to record the temperature of individual fibers by embedding a liquid crystal that changes color. This is desirable for measuring such stress-related phenomena as creep, when plastic deformation is accompanied by a local temperature rise.

*Sensored Composites — copyright and patent, Richard J. Weiss, 1990

The usual procedure in dealing with the patent office found the inventor on a merry-go-round. The patent office invariably cites other patents that they believe already cover the invention. Eventually persistence is rewarded and a good attorney earns his fee when the patent office drops its objections. When this good news reached Richard he was also notified of the bad news. The

Air Force had clamped a security stamp on the patent — the idea must be developed by Americans and in the interim no patent number would be granted. Thus the details of the patent had to be kept under wraps.

Richard spent several years trying to take the idea the next step but has had no response. Why the security clamp? It is a device to prevent foreign powers from exploiting American ideas and technology. The cold war was over but government red tape tenaciously stuck to everything it touched. In desperation Richard wrote Vice President Al Gore and to Secretary of the Air Force Sheila Widnall but received no response. In the interim he believed he was in worse shape if he had never applied for a patent since he would not have been forbidden from publishing details at a scientific conference. He confided his plight to a patent consultant at IBM who advised Richard to sit on the 'phantom' patent until some big company developed the idea and then sue them. Is that the way to run an airline?

The BBC monologue that Richard wrote began with a short musical passage on the glass harmonica. The story of this instrument dated back to 1762 when Franklin first heard the musical glasses in Italy. These consisted of wine glasses filled with water to varying levels such that each sounded a different note when the player's wetted finger rubbed the rim. The logistics were awkward, requiring frequent adjustment of the water levels to keep the glasses in tune and presenting manipulative difficulty in exciting chords of notes. Professor Charles Taylor at the University of Cardiff in Wales first brought the tale to Richard's attention in the early 80's. Taylor had recorded a tape for the BBC that analyzed the physics of various musical instruments. He had found an original glass harmonica at the Royal College of Music and had it repaired to produce the tape.

Franklin had altered the basic design of water-filled glasses by having hollow glass hemispheres blown, tuned by grinding, and then mounted on a horizontal shaft that was spun with a foot treadle. The player set the notes ringing by touching the rim of the appropriate glass with a wetted finger. The sound was pleasant and unique, so much so that Mozart, Beethoven, and Gluck composed a few pieces for the instrument.

When a quack Anton Mesmer began using the instrument in Paris as background music to hypnotize his patients, rumors were spread that the instrument could drive one insane. It was eventually outlawed in Germany, in spite of Lafayette's interest in its therapeutic value. Except for Professor Taylor's interest in resurrecting the instrument it had remained virtually unknown for almost two centuries. In the early 1980's Richard discussed its musical qualities and its construction problems with a musician he met at Blanchard's Tavern. Richard was referred to a scientific glass blower in Waltham, Mass., Gerhard Finkenbeiner, himself an amateur musician. Gerhard, a 5'6" likeable man with a perpetual smile had lived in France before emigrating to Massachusetts, but he took to the challenge instantly. Charles Taylor had claimed that the original instruments suffered from the use of poor quality glass. Actually this was probably not the reason but Gerhard fabricated the glass hemispheres from pure silica (glass with no added impurities) and produced a playable instrument. His shop on Rumford Ave. in Waltham, Mass. began to fabricate and market the instrument. Eventually it attained worldwide recognition.

By setting up a stroboscope one could 'freeze' the oscillatory modes of vibration. The round shape of the quiescent glass hemispheres developed oval shaped mode distortions that set the air in vibration. The fundamental tone dominates the emitted tone producing a sound reminiscent of clinking crystal glass.

A few years passed and Richard was invited to a bicentennial celebration in Franklin, Mass, a town named after Benjamin Franklin while he was American Ambassador to France during the 1780's. The town fathers asked the famous Ambassador to contribute money to purchase a bell for the town. Franklin responded by thanking them for naming the place after him but suggested that he should contribute books rather than a bell. "I much prefer sense to sound."

Franklin asked a friend to select about 100 books amongst the classics. These were sent to Franklin, Massachusetts and provided the basis for the first free lending library in America. In 1900 money was raised to build a new library along classic Greek lines in order to display the Franklin gift. An Italian muralist was invited to add appropriate scenes to the bare walls. In keeping with the architectural style nude mythological goddesses prancing in the outdoors were painted and an inaugural celebration arranged to dedicate the new facility. When the wives of the town officials first saw this example of 'immoral art' they pressured their husbands to bring in a second artist to add veils over the goddess' privates.

Richard attended the bicentennial ceremony in tavern costume and read from 'Poor Richard's Almanac'. This was followed by a most talented young lady Alisa Nakashian playing the glass harmonica and singing appropriate tunes. Her silky, clear voice so enchanted Richard that he wrote a monologue for Sally Franklin in which she reminisced about her life with her famous father and sang the Scottish songs he loved so well. This did so well that Alisa was invited to perform at First Night in Boston. Since then she has appeared at various venues doing her act, including performances at a theater in Cambridge where a second monologue of Richard's (Kaiser Bill) rounded out the evening.

Richard's fascination with German history for the last century began with a suggestion made by Seweryn Chomet, his colleague at King's College London, who had unearthed some of the details of a Royal scandal in 1875 when the Churchills and the Royal family were at loggerheads over an intrigue between Blandford Churchill (the young Winston's uncle) and a Lady Aylesford, a woman who had also captured the attention of the Prince of Wales (later Edward VII). It required the diplomatic intervention of Disraeli to settle it to Queen Victoria's satisfaction and keep it out of the press. Richard converted the story into a novel but in the course of his research the interplay between members of the Royal family and their marriages intrigued him. Kaiser Wilhelm II, Victoria's grandson, and his hatred for his mother and his uncle made for interesting reading.

'Willy', as his uncle King Edward called him, was born with a lame left arm and this affected his personality, was cause for his hatred for his mother who had forced him into undergoing a painful but unsuccessful cure, and his dislike for uncle Bertie who refused to address him as Kaiser Wilhelm. Nonetheless, Wilhelm developed an outgoing personality to mask his lameness and he would constantly demand everyone's attention. He frequently recited Kipling, gave long speeches, sang "When I Was a Lad" to Arthur Sullivan who exclaimed it was the first time the song was sung by a real admiral, and often appeared in public in military dress. Hence the idea occurred to Richard to bring Kaiser Bill out of exile in 1923 to secretly perform in an English music hall. Singing songs of the period, reminiscing about his hated mother and uncle, professing his respect for Queen Victoria, and proclaiming his innocence over the start of WWI. The monologue achieved success, eventually running in a theater near Boston for a month.

When Hitler ordered the invasion of Holland on May 10, 1940 Winston Churchill suggested to King George VI that an RAF plane be flown to Doorn, the site of the Kaiser's exile. Churchill contended that WWI was not the Kaiser's fault, it was his fate. The King agreed with Churchill and an emissary was sent to the Kaiser with the offer. Wilhelm, over 80, decided to remain in Holland and accept whatever fate Hitler had in store for him. But Hitler had ordered his troops to ignore the Kaiser. Wilhelm died in 1941, the day before Hitler invaded Russia.

The fascination Richard held for the Kaiser's story brought him into focus with the Lusitania story, a Cunard passenger liner that was sunk by a U boat in April 1915, an event that exacerbated American hatred for the Hun. Richard tuned it into a spy novel as seen from the German point of view. The most fascinating aspect of WWI was the intense hatred that developed and the blind military campaigning that led to such unwarranted and massive bloodshed. This foisted on the Germans the harsh terms of the Versailles Treaty, the root cause for the rise of the Third Reich and WWII.

In December 1993 Richard approached his 70th birthday. He had visited over 50 laboratories and had produced popularized accounts of the work for magazines and newspapers. While the challenge of transcribing these research efforts into easily readable articles kept him on his toes the most exciting task in combining science and writing was initiated by his own brother Sanford who enlisted Richard's help to authenticate several Picasso oil paintings. The story is summarized in the press release prepared by Richard, specifically written to make a dramatic impact:

FAMOUS PICASSO DRAWING CHALLENGED
Science and Art Experts Ready to Battle
The most extensively reproduced Picasso drawing *DON QUIXOTE AND SANCHO PANZA*, signed and dated 10.8.55 (August 10, 1955), has been scientifically determined to be a copy of an identical oil painting by Picasso, signed and dated 10.3.47 (March 10, 1947). This discovery of an earlier work has the scientific and art experts prepared for open battle.

The story of this astounding find takes us back to the Spanish Civil War (1936–1939) when Vincent Polo, a Loyalist fighter pilot ferrying planes from France to Spain, was introduced by the Spanish Ambassador to the self-exiled Picasso at the Embassy in Paris. Picasso entrusted Polo to fly a sealed package (probably money) to Picasso's mother in Barcelona. Following the delivery, Picasso extended an invitation to Polo to visit him whenever he was in France. Later captured by Franco's forces, Polo escaped to America and became a citizen, serving in the American Forces during WWII.

It wasn't until 1947 that Polo, his wife and new son, traveling in Europe before taking up a position in Mozambique were able to visit Picasso at his studio in Golfe-Juan on the French Riviera. Picasso expressed his thanks for the previous favor by preparing for Vincent Polo the now-famous (24" × 28") windmill-slashing Don Quixote theme to commemorate the 400th anniversary of the birth of Cervantes (b 1547) and to symbolize the overwhelming odds faced by the Loyalist Air Force against Franco's forces. To personalize the oil on canvas painting Picasso included the names Ann and Vincent Polo in the foreground, copying the signatures from a note Polo left for Picasso. The painting was dated 10.3.47 (March 10, 1947). In 1955 Picasso prepared a scaled down (14" × 16") drawing that he presented to his friend Pierre Daix, Paris publisher and art critic, changing the date and making less distinct the name Polo. It is this drawing, now at the Museum in St. Denis, north of Paris, that has been liberally reproduced for collectors around the world.

Polo returned to America and served 24 years in the Brooklyn College and New York University Engineering Departments. Suffering from emphysema in 1979 Polo approached Sotheby's to auction the works for his wife and son but, not having a signed bill of sale, the auctioneers refused to handle the works. They cautioned that Picasso was a popular target for forgers.

Fred Polo, in going over his father's effects in 1986, found a series of Kodachrome slides taken on the occasion of his own Holy Communion in 1954. One of these slides clearly revealed the Don Quixote painting hanging on the living room wall. This Kodachrome was taken at least a year before the smaller drawing was given to Daix for publication.

SCIENCE TAKES OVER

Dr. William Croft, physicist at the materials Research Laboratory in Watertown, corroborated the date of the Kodak-processed slide from the mount, changed early in 1955. Hence the Polo Kodachrome proved the existence of the painting before the drawing.

The McCrone Research Institute, having worked on dating the famous Turin Shroud, was enlisted to establish the authenticity of the Polo painting. Its findings, including carbon dating, clearly verify the pre-1955 date for the Polo painting.

ART EXPERT DISPUTES FINDING

The Institute For Art Research in New York was also enlisted in an effort to establish the authenticity of the Polo painting. They submitted it to John Richardson, well-known Picasso expert. Richardson declared the painting to be a forgery.

How did Richard get involved in this controversy? Fred Polo had approached Sanford to help in the authentification and Sanford asked Richard for his scientific help. Richard found several experts and the corroborating scientific clues were outlined in the letters from Drs. Croft and McCrone:

Mr. John Richardson
186 E. 75 Street
New York, NY 10021   Oct. 10, 1991

Dear Mr. Richardson,

I have examined four Kodachrome slides of Mr. Polo and compared the cardboard mount with my own extensive collection of slides, which have been processed by the Kodak Company of Rochester, NY.

I can verify that the mounting format of Mr. Polo's is consistent with the period in my work from 1948 to Mar 1955 after which the format changed. Copies of slides prepared by Kodak are included for reference.

It would seem to me that the Don Quixote painting, partially observable in one of the slides, existed in Mr. Polo's collection prior to the well known lithograph by Picasso dated 10/3/1955.

I understand that this mounting format was used only by Kodak. Mounts available for do-it-yourself workers were quite different in format.

Sincerely yours
s/ William J. Croft

McCrone Research Institute
2820 South Michigan Avenue
Chicago, Illinois 60616-3292 USA
Fax (312) 842-1078
Phone (312) 842-7100

15 June 1993

Mr. Sanford Weiss
Golden State Financial
4717 Van Nuys Blvd.
Sherman Oaks, CA 91403

Dear Mr. Weiss:

I would like to summarize the basis for my conclusion regarding "El Quijote de la Mancha" an oil-on-canvas painting measuring 23-7/8" × 27-7/8". The provenance evidence you have so carefully developed is especially convincing in showing that Picasso painted Don Quixote in 1947 as a gift to Vincent Polo.

The videotaped deposition of Mr. Polo in 1985 certainly rings true in establishing a reasonable explanation for Picasso's generosity toward Mr. Polo. I was particularly impressed by Picasso's working of Vincent Polo's signature and his wife's name (Ann) into the foreground of Don Quixote. I've appended a copy of Mr. Polo's signature to compare with the foreground of the Don Quixote; the resemblance is obvious. This I find very convincing. It is equivalent to Picasso writing "To my friend Vincent Polo," then signing it with the date.

The date on the painting (March 10, 1947) agrees with Mr. Polo's account of his travels at that time: England, France, South Africa and with Picasso's home and studio location. Polo's signature was taken from a certificate of inoculation against yellow fever he signed on 26 November 1947 in London.

Almost equally impressive, is the color transparency showing the Don Quixote painting and Mr. Polo's young son taken in 1954, therefore proving the painting existed at that time. The

Kodachrome transparency, furthermore, has the typography style consistent with a date early in 1955. That style changed in 1955. I was able to confirm the fact the Polo transparency had to have been processed by Kodak before March of 1955 by reference to my personal collection of slides dated over the period 1945-today. You also obtained a similar comparison and an identical conclusion from a William Croft who added the dates for the two styles of transparency mount. Kodak last processed transparencies in the earlier mounts in March 1955. This very conclusively shows a photograph to prove the Polo-Don Quixote predates August 1955.

I note that M. Pierre Daix has no solid evidence to support his contention that Picasso rendered the Don Quixote drawing for him in August 1955 but, in any case, his dates on the drawing are considerably later than the now well-substantiated 1947 date for the Polo-Don Quixote.

Further substantiating evidence includes carbon dating of the Polo-Don Quixote canvas. The date turns out to be "prebomb" which means before 1955. Although the usual error in carbon is ±50 to ±100 years the effect of atmospheric nuclear testing which began in 1954 makes it possible to be certain the Polo-Don Quixote canvas is pre-1955.

The pigment data are consistent with Picasso's palette during the 1947 period. The painting, black on white, shows lead white and whiting (calcium carbonate) in the background and bone black with ultramarine for the figures. The addition of blue pigment seems to be characteristic of Picasso's blacks, at least at that time. The "Charnel House," for example, from the same period shows blue with black pigment used.

A significant portion of my career has been spent as a criminalist on variety of civil and criminal cases. Although the Polo-Don Quixote is not a legal matter, the scientific approach and reasoning used to decide the question: — Was this Don Quixote painted by Picasso in 1947 and therefore the original of which the Saint Denis drawing is a later Picasso version? — is no different. I would feel very comfortable and confident if I were an expert witness in court defending a positive answer to the above question. The evidence as cited in this letter is, in fact, more definite and more conclusive than criminalists usually have in support of their conclusions.

The defense now rests its case, Your Honor.

Sincerely,

Walter C. McCrone

WCM:nbd

Adding to the usual varieties of singers, etc. at the tavern Richard arranged for special performances by actors portraying Calvin Coolidge, Daniel Webster, Theodore Roosevelt, Amelia Earhart, Kaiser Bill, Sally Franklin, and Louisa May Alcott. Hopefully this effort entitled 'Recalled to Life' will continue to provide realistic encounters of famous historical personages with modern audiences.

Richard Weiss
Dec. 14, 1993

INDEX